Fast Track to Differential Equations

Albert Fässler

Fast Track to Differential Equations

Applications-Oriented—Comprehensible—Compact

Second Edition

 Springer

Albert Fässler
Bern University of Applied Sciences
Biel/Bienne, Switzerland

ISBN 978-3-030-83452-4   ISBN 978-3-030-83450-0   (eBook)
https://doi.org/10.1007/978-3-030-83450-0

Preface to the German Original

> *A thick book is*
> *a great evil.*
>
> *Gotthold Ephraim Lessing*

With this book I honor posthumously Dr. Dr. h. c. mult. Eduard Stiefel, late professor of Mathematics at ETH Zurich, Switzerland. He shaped my mathematical thinking sustainably and promoted me as his teaching assistant and later as his research assistant and lecturer. Prof. Stiefel had an excellent international reputation in the professional world. He was also a computer pioneer and a highly esteemed teacher.

Usually, a book on differential equations deals with computing solutions by means of some special technical knowledge, such as complex numbers or partial fraction decompositions. For many students not majoring in mathematics and at a time when even calculators can integrate symbolically, computing solutions has become an unnecessary annoyance that just impedes the access to the topic.

Here, however, the focus throughout is on more complex problems and on verifying the correctness of given solutions by differentiating and plugging them into the equations.

This includes several advantages:

(a) Verifying the correctness promotes an in-depth understanding[1].
(b) Differentiation is covered in high school already.
(c) The time required is much shorter.
(d) The time gained is used for mathematical modeling (that is, for establishing differential equations of a given problem) and for interpreting solutions, often by generating their graphs.

The rather lengthy and often cumbersome technical proofs of existence theorems have been deliberately omitted in favor of a diverse range of applications from different fields. Also they are available on demand in the literature and on the Internet. On the other hand, I have laid emphasis on counterexamples and, of course, on the knowledge of the content of such theorems.

In the beginning we limit ourselves to an initial understanding of simple examples of differential equations, namely of those that can be solved by hand with elementary

---

[1] The Hungarian mathematician George Polya (1887-1985) strongly supports this postulate in [32].

integrals or to cases where the resulting integrals can be solved by use of a calculator or a computer.

For more complex problems, solutions are furnished, not least because computer algebra systems implemented on computers or calculators can often solve differential equations.

Of course, numerical methods come into play when integrals are no longer solvable in closed elementary form. This often happens in more demanding problems. Then graphs are particularly useful.

By way of examples, individual commands and instructive short programs (with explanations) of the computer algebra system "Mathematica" with the corresponding outputs are provided. A more convenient alternative to using a calculator is to enter the commands for calculations and plots directly on the website of "Wolfram Alpha".

The book is aimed at a broad spectrum of interested parties:

- Ranging from high school students at the end of their upper secondary education (cantonal schools) who are highly motivated in mathematics and the sciences
- through college and university students majoring in engineering, in the natural and computer sciences, and in economics,
- and to lecturers at colleges of education teaching subject-oriented classes,
- all the way through to mathematics and physics students in their first semesters who want to tackle the topic in an unceremonious and application-oriented way.

In selecting topics for investigation, emphasis was put on analyzing problems from the natural sciences and on playful problems.[2] After all, playfulness and aesthetics can be cultivated in mathematics and geometry. I have given space to both aspects. These together with novel challenging puzzles, along with my professional intent, have motivated me to study mathematics.

It was probably no coincidence that a course, which I gave only a few years ago, with this endeavor was appreciated by the students at the College of Education of the University of Applied Sciences of Northwestern Switzerland.

Of course, mathematics is also used on a broad basis as a tool for modeling and solving scientific problems. Many years of experience as a lecturer at the Department of Technology and Computer Science at the Bern University of Applied Sciences BFH have been drawn into here.

Chapter 1 offers a refresher of the calculus required including many applications-related examples and exercises. Meanwhile, I have refrained from the tedious and boring introduction of the exponential function via fractional exponents in favor of an introduction via power series by means of a directional field. This is already a gentle introduction to the special differential equation $y' = y$, made with an onset for a sequel!

---

[2] Sign in a roundabout in Bhopal, India: "Play is the brain's favorite way of learning".
Friedrich Schiller: "Man is only human when he plays."

I have also tried to intersperse problems that are fun.

For further reading one may consult [4].

Words of Thanks

A very special big thank-you goes to my fellow student, Dr. Baoswan Dzung Wong. She has contributed both in content and editorially to a significant improvement of the book.

My valued colleague Dr. Walter Businger at the Department of Technology and Computer Science of the Bern University of Applied Sciences (BFH) contributed to the diagrams and always took time to solve my LaTeX problems. Joint mathematical activities and technical discussions connect me with him for many years.

The reference to [21] by Prof. Dr. Daniel Farinotti, glaciologist at ETH Zürich, inspired my contribution on *Global Warming*.

For the pleasant cooperation with Dr. Annika Denkert (editorial), Anja Groth and Barbara Lühker (project management) from Springer Publishing and Tatjana Strasser (copy-editing) a thank-you goes to Heidelberg.

Many thanks to the ladies of the library of the Department of Technology and Computer Science at the BFH in Biel/Bienne for their help.

Finally, I thank Olivier Fässler for solving my computer problems and, last but not least, my wife Carmen Fässler for creating a quiet and comfortable work environment and for her understanding when I spent many hours on my computer.

I wish you a pleasant journey on your fast track to differential equations!

CH-2533 Evilard, September 2017
Albert Fässler

Preface to the English Translation

With Dr. Baoswan Dzung Wong I had the pleasure to study at ETH Zürich. After-wards she continued her studies at the University of California, Berkeley, U.S.A. and finished with a Ph.D. in the field of functional analysis under the guidance of Pro-fessor Tosio Kato. With her precise work style and her highly professional abilities, she improved this edition from the linguistic and mathematical point of view. I wish to express my big thanks to her for the beautiful work and the inspiring interaction during the process of realizing this book.

Thanks go also to Nicolas Gruber, Professor of Environmental Physics at the Insti-tute of Biogeochemistry and Pollutant Dynamics at ETH Zurich, for his support in the section on *Climate Change*.

The two sections on *Kalman Filter, Brownian Motion and Langevin Equations* are contributed by Dr. sc. techn. ETH Dacfey Dzung. I thank him for this and also for proofreading parts of the manuscript.

For implementing the layout of this book, an extended version of the German origi-nal entitled *Schnelleinstieg Differentialgleichungen*, I wish to express my thanks to Mr. Sudhany Karthick in Chennai, India.

Finally a statement by Dr. Alessio Figalli, Professor at ETH, Zurich and Fields medalist 2018:
"I am happy to see such a book. It will serve as a support for many students, pro-fessors and faculty."

CH-2533 Evilard, October 2019
Albert Fässler

Preface to the Second Edition

Chapter 6 is new:

6.1 Climate Change.
This section deals with a simple mathematical model for analyzing the future temperature profile of the earth's surface in relation to the carbon dioxide problem.

6.2 Epidemiology.
Analysis of the SIR and SEIR models taking into account the COVID-19 pandemic.

And the following problems (with their numbers) are new:

22 Geometry of Railway Tracks, Roads and Ski Jumps,
29 Pedal Curve,
31 Brainteaser: The Rain Problem,
65 Tree Growth According to Chapman-Richards,
97 Climate Data,
98 Carbon Budget for the 1.5 Degree Target,
99 Time Dependent Radiation Function g(t),
100 SIR Model,
101 Vaccination Campaign.
102 SEIR Model.
103 Second Wave.

I would like to thank Professor Dr. Nicolas Gruber, Professor of Environmental Physics at the Department of Environmental Systems Science at ETH Zurich, for his extremely valuable and active contribution to the topic of climate change.

My thanks also go to Dr. sc. techn. ETH Dacfey Dzung for his support in the section on epidemiology.

I again had the privilege to have Dr. Baoswan Dzung Wong as a translator from German into English and for proofreading. I owe her many thanks.

A *merci* goes from Western Switzerland to Heidelberg for the renewed pleasant collaboration with Dr. Annika Denkert (editing) and Ms. Anja Groth from Springer Verlag, regarding the second edition entitled *Schnelleinstieg Differentialgleichungen*.

I am grateful to my brother Markus Fässler. He left his alternative summer residence on Lake Constance to me to work on the epidemiology section. It was an ideal place in the face of the coronavirus situation in the summer of 2020.

CH-2533 Evilard, June 2021
Albert Fässler

Contents

Chapter 1
Prerequisites from Calculus

1.1 Exponential and Logarithm Functions

1.1.1 Exponential Function exp(x) as a Power Series

Let us try to find a function f with the following two properties:
For all $x \in \mathbf{R}$

$$f'(x) = f(x) \tag{1.1}$$
$$f(0) = C \tag{1.2}$$

(a) First we approach this problem geometrically by using a so-called **direction field**. The idea is to draw a large number of "compass needles" aligned according to the slopes given at the y-values where $y = f(x)$. The horizontal lines are so-called **isoclines**, i.e. lines along which the slope $f'(x)$ is constant,

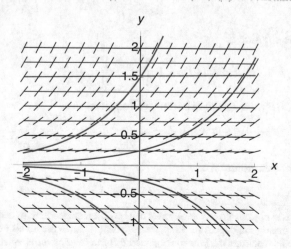

(b) Evidently, for each value of C, there exists a solution to the problem (1.1), (1.2).

(c) For the special case $C = 0$, the solution is $f(x) = 0$ for all $x \in \mathbf{R}$.

Polynomials cannot be solutions, because after differentiation a polynomial of degree n becomes one of degree $(n-1)$.

We try a formal approach by so-called **power series:**

$$f(x) = a_0 + a_1 x + a_2 x^2 + a_3 x^3 + \ldots = \sum_{i=0}^{\infty} a_i x^i. \tag{1.3}$$

The derivative is[1]

$$f'(x) = a_1 + 2a_2 x + 3a_3 x^2 + \ldots = \sum_{i=1}^{\infty} i a_i x^{i-1}. \tag{1.4}$$

Comparing coefficients of (1.3) and (1.4) yields the requirements

$$a_1 = a_0 \qquad a_2 = \frac{1}{2} a_1 \qquad a_3 = \frac{1}{3} a_2 \qquad \ldots$$

or more generally,

$$a_{k+1} = \frac{1}{k+1} a_k.$$

The initial condition (1.2) implies $f(0) = a_0 = C$.

Consequently, the coefficients a_k are obtained recursively:

$$a_0 = a_1 = C \qquad a_2 = \frac{C}{2} \qquad a_3 = \frac{C}{2 \cdot 3} \qquad \ldots$$

or more generally,

$$a_k = \frac{C}{k!}.$$

Therefore, the solution is

$$f(x) = C \cdot \left(1 + x + \frac{1}{2!} x^2 + \frac{1}{3!} x^3 + \frac{1}{4!} x^4 \ldots \right).$$

Definition 1. *By use of the* **exponential function**

$$\exp(x) = 1 + x + \frac{1}{2!} x^2 + \frac{1}{3!} x^3 + \frac{1}{4!} x^4 \ldots = \sum_{k=0}^{\infty} \frac{x^k}{k!}$$

we may rewrite the solution as

$$f(x) = C \cdot \exp(x).$$

---

[1] It is by no means obvious that a power series can be differentiated term by term. A more precise analysis shows that we may rightfully do so, because $\exp(x)$ converges for all $x \in \mathbf{R}$ (radius of convergence $= \infty$) and therefore, the function obtained by term-by-term differentiation also converges for all $x \in \mathbf{R}$.

Nevertheless, we have to prove the **convergence** of our power series.

Proof. First let $x \geq 0$ be a fixed, but arbitrary number. We choose m large enough such that $q = x/m < 1$.

If we truncate the power series after the first m summands, we may estimate the remainder as follows:

$$\frac{x^m}{m!} + \frac{x^{m+1}}{(m+1)!} + \ldots = \frac{x^m}{m!}\left(1 + \frac{x}{m+1} + \frac{x^2}{(m+1)(m+2)} + \ldots\right) \leq \frac{x^m}{m!}(1 + q + q^2 + \ldots)$$

The geometric series $1 + q + q^2 + q^3 + \ldots$ is known to converge to $\frac{1}{1-q}$ if $|q| < 1$. Therefore, the strictly monotonically increasing power series of $\exp(x)$ is bounded for $x \geq 0$ and hence converges.

The convergence for $x < 0$ follows from the fact that every alternating series with $\lim\limits_{k \to \infty} a_k = 0$ converges.   q.e.d.

Let us summarize our main result and make it more precise in the following

Theorem 2.   *There is exactly one function f satisfying the two properties that for all $x \in \mathbf{R}$*

$$\left.\begin{array}{l} f'(x) = f(x) \\ f(0) = C \end{array}\right\},$$

namely $f(x) = C\exp(x)$.

It remains to prove that the solution is unique.

Proof. Let us assume that other than $f(x) = C\exp(x)$ there is a function g satisfying the two properties. Now we consider the derivative of $\frac{g}{f}$.

As $f' = f$ and $g' = g$, we obtain by use of the Quotient Rule

$$\left(\frac{g}{f}\right)' = \frac{g'f - gf'}{f^2} - 0.$$

Hence $\frac{g}{f} = k = \text{const} \Rightarrow g = k \cdot f$.   Because $f(0) = g(0) \Rightarrow f = g$.   q.e.d.

The following example shows that the question of uniqueness cannot be answered solely by means of the direction field.

Example 1.1   The problem

$$\left.\begin{array}{l} f'(x) = \sqrt{f(x)} \\ f(0) = 0 \end{array}\right\}$$

has infinitely many solutions (C being an arbitrary constant), namely

$$f_1(x) = 0 \quad \text{and} \quad f_2(x) = \begin{cases} 0 & \text{if } x \leq C \\ \frac{1}{4}(x - C)^2 & \text{if } x > C \end{cases}.$$ ◇

Definition 3.   *The function value*

$$e = \exp(1) = 1 + 1 + \frac{1}{2!} + \frac{1}{3!} + \ldots = \sum_{i=0}^{\infty} \frac{1}{i!} = 2.718281828459045235\ldots \notin \mathbf{Q}$$

*is called **Euler's number**.*

It is irrational.

1.1.2 Properties of exp(x)

$$\exp'(x) = \exp(x), \qquad \exp(0) = 1, \qquad \exp(x) \geq 1 \quad \forall x \geq 0.$$

The **Addition Theorem of the Exponential Function**, which is important and entails further properties, states that

$$\exp(a) \cdot \exp(b) = \exp(a+b). \tag{1.5}$$

Proof. Expanding the product of the two series and collecting all terms of the form $a^k b^{n-k}$ for $k \leq n$, n fixed, yields

$$\exp a \cdot \exp b = \left(1 + a + \frac{a^2}{2!} + \ldots + \frac{a^n}{n!} + \ldots\right) \cdot \left(1 + b + \frac{b^2}{2!} + \ldots + \frac{b^n}{n!} + \ldots\right)$$

$$= 1 + (a+b) + \left(\frac{a^2}{2!} + \frac{b^2}{2!} + ab\right) + \left(\frac{a^3}{3!} + \frac{b^3}{3!} + \frac{ab^2}{2!} + \frac{a^2 b}{2!}\right) + \ldots$$

The third term equals $\frac{(a+b)^2}{2!}$, the fourth term equals $\frac{a^3 + 3a^2 b + 3ab^2 + b^3}{3!} = \frac{(a+b)^3}{3!}$.

In order to obtain the general $(n+1)$st term, we collect for fixed n all powers $a^k \cdot b^\ell$ with $k + \ell = n$:

$$\frac{a^n}{n!} + \frac{n}{n!} a^{n-1} b + \frac{n(n-1)}{n! \cdot 2!} a^{n-2} b^2 + \frac{n(n-1)(n-2)}{n! \cdot 3!} \cdot a^{n-3} b^3 + \ldots + \frac{b^n}{n!} = \frac{(a+b)^n}{n!}.$$

Hence

$$\exp a \cdot \exp b = 1 + \frac{a+b}{1!} + \frac{(a+b)^2}{2!} + \ldots + \frac{(a+b)^n}{n!} + \ldots = \exp(a+b).$$

q.e.d.

In particular, we have

$$\exp(x) \cdot \exp(-x) = 1 \quad \Longleftrightarrow \quad \exp(-x) = \frac{1}{\exp(x)}, \tag{1.6}$$

whence $\qquad 0 < \exp(x) < 1 \; \forall x < 0 \quad$ and $\quad \lim_{x \to -\infty} \exp(x) = 0. \tag{1.7}$

By inspecting the direction field of $\exp(x)$ we find, rather surprisingly, that for varying C the graphs of the functions $C \cdot \exp(x)$ are mutually congruent. They are carried into each other by horizontal shifts.

Repeated use of the Addition Theorem yields
$$\exp(n) = \exp(\underbrace{1 + 1 + \ldots + 1}_{n \text{ summands}}) = [\exp(1)]^n.$$

Hence, with $e = \exp(1)$, we can also write $\exp(n) = e^n$ where $n \in \mathbf{N}$.

Likewise for $m \in \mathbf{N}$,

$$e = \exp(1) = \exp(\underbrace{1/m + 1/m + \ldots + 1/m}_{m \text{ summands}}) = \left[\exp(\tfrac{1}{m})\right]^m.$$

Therefore, $\exp\left(\frac{1}{m}\right) = e^{\frac{1}{m}}$ where $m \in \mathbf{N}$.

Furthermore, for $p, q \in \mathbf{N}$

$$\exp(\frac{p}{q}) = \exp(\underbrace{1/q + 1/q + \ldots + 1/q}_{p \text{ summands}}) = \left[\exp(\tfrac{1}{q})\right]^p = \left[e^{1/q}\right]^p = e^{p/q}.$$

With $r = \frac{p}{q} > 0$, we have $\qquad \exp(r) = e^r$ where $r \in \mathbf{Q}$,

and thus $\qquad\qquad\qquad\qquad \exp(-r) = \dfrac{1}{\exp(r)} = e^{-r}.$

Since the rational numbers are dense on the number line (that is, each arbitrarily small interval contains infinitely many rational numbers), we are not in conflict with anything known or previously defined when writing for **all real numbers** x,

$$e^x = \exp(x). \tag{1.8}$$

In the literature, both notations are used. Hence we may also write the Addition Theorem as follows:

$$e^a \cdot e^b = e^{a+b}.$$

1.1.3 Exponential Function

Because $(e^{kx})' = k \cdot e^{kx}$, we may generalize Theorem 2:

Theorem 4.

$$\left.\begin{array}{l} f'(x) = k \cdot f(x) \\ f(0) = C \end{array}\right\} \Longleftrightarrow f(x) = C \cdot e^{kx}$$

The following figure shows the graphs of $e^{\pm 1x}$, $e^{\pm 1x/2}$.

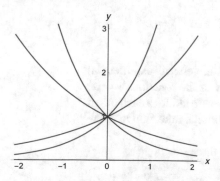

For $k > 0$, the function e^{kx} is strictly monotonically increasing.
For $k < 0$, the function e^{kx} is strictly monotonically decreasing.

The graphs are mutually **affine**, because they are carried into each other by scaling, i.e. by linear stretching or compressing in x-direction. If the k-values have different signs, an additional reflection in the y-axis is required.

Let $f(x) = Ce^{kx}$ be the general exponential function, and let us consider the sequence of function values $f(x_n)$, where x_n are equidistant points with step size $h > 0$:

$$f(x_n) \quad \text{with} \quad x_n = x_0 + k \cdot h, \quad k \in \mathbf{Z}.$$

Comparing consecutive function values at $x_{n+1} = x_n + h$ and using the Addition Theorem yields

$$f(x_{n+1}) = C \cdot e^{k(x_n+h)} = C \cdot e^{kh} e^{kx_n} = e^{kh} \cdot f(x_n).$$

The factor $= e^{kh}$ does not depend on the choice of x_0.

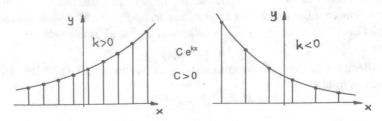

Exponential growth or exponential decay means multiplicative behavior with factor $\exp(kh)$ for every discrete sequence of function values with equidistant step size $h > 0$ (see figures). In fact, the sequences are geometric.

1.1.4 Change of Base and Hyperbolic Functions

The function a^x (where $a > 0$) can always be rewritten in the base e, because $a = e^{\ln a}$ and, therefore,

$$a^x = (e^{\ln a})^x = e^{(\ln a) \cdot x} = e^{kx} \quad \text{where } k = \ln a.$$

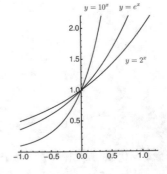

Hence e^{kx} is called the **general exponential function**. The graphs of the functions $y = a^x$ where $a > 1$ and $y = e^x$ are affine to each other. For $0 < a < 1$, an additional reflection in the y-axis is required.
The figure shows the three functions

$$e^x, \quad 2^x = e^{(\ln 2) \cdot x}, \text{ and } 10^x = e^{(\ln 10) \cdot x}.$$

Definition 5.

$$\cosh x = \frac{e^x + e^{-x}}{2} \qquad \sinh x = \frac{e^x - e^{-x}}{2}$$

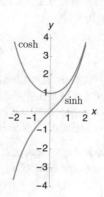

are called the hyperbolic cosine and the hyperbolic sine functions.

Except for the signs, hyperbolic identities resemble trigonometric identities.

Example 1.2 As is easily verified, the following holds:

$$(\cosh x)' = \sinh x, \qquad (\sinh x)' = \cosh x,$$

$$\cosh^2 x - \sinh^2 x = \left(\frac{e^x + e^{-x}}{2}\right)^2 - \left(\frac{e^x - e^{-x}}{2}\right)^2 = 1. \qquad \diamond$$

1.1.5 Logarithm Function

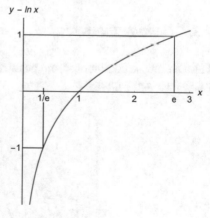

Definition 6. *The inverse of the exponential function e^x is called $\ln x$ (natural logarithm).*

Therefore,
$$y = e^x \iff x = \ln y.$$
A word of caution: The logarithm function is only defined for positive variables!

Obviously,

$$e^{\ln(a)} = a \text{ for all } a > 0 \qquad \text{and} \quad \ln(e^b) = b \text{ for all } b \in \mathbf{R}.$$

As in the case of exponential functions, we may perform a **base change** with logarithm functions:

Definition 7.
$$y = b^x \iff x = \log_b y \qquad (b > 0).$$

The equation $y = b^x = e^{\ln b \cdot x}$ implies $\log_b y = x$ and $\ln y = \ln b \cdot x$.
Substituting x from the left-hand equation into the right-hand equation yields the simple relation

$$\log_b y = \frac{1}{\ln b} \cdot \ln y \qquad \text{or} \qquad \log_b x = \frac{1}{\ln b} \cdot \ln x.$$

The various logarithm functions differ only by a multiplicative constant. Their graphs are mutually affine with stretching or compressing in vertical direction. This is not surprising, as it corresponds to the affinity of the various exponential functions b^x in horizontal direction.

Practical consequence: On the pocket calculator the key for the natural logarithm ln suffices for computing any $\log_b$.

Caution:

- Often the natural logarithm ln is denoted log.
- Sometimes log or lg is used for the decimal $\log_{10}$.
- In computer science it is customary to use lb or ld for the binary logarithm $\log_2$.

1.2 Integral Calculus

1.2.1 Definite Integral

Let a piecewise continuous[2] and positive function $f(x) \geq 0$ be given, as well as the two values a, b with $a < b$.

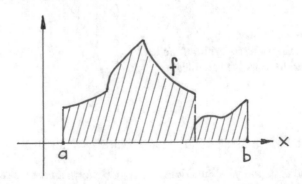

Graphically, the area of the shaded region corresponds to the **definite integral** of the function $f(x)$ with integration limits a and b.

---

[2] A piecewise continuous function is continuous except for possible isolated points of discontinuity. Continuity comprises piecewise continuity.

Notation:

$$\int\limits_{a}^{b} f(x) \cdot dx.$$

In order to generalize the above, the definite integral of a function with negative values is defined to have a negative area. As in the beginning, we assume that the integration limits $a < b$, that is, integration is performed from left to right:

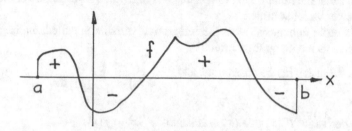

If integration is performed from right to left, then the sign changes.

Example 1.3

$$\int\limits_{0}^{2\pi} \sin x \cdot dx = 0.$$

$\Diamond$

The **Mean Value Theorem for Integrals** states that for a continuous function f, there exists at least one intermediate number $z \in (a,b)$ such that

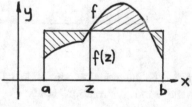

$$\int\limits_{a}^{b} f(x) \cdot dx = f(z) \cdot (b-a).$$

This is clear in an intuitive sense: The rectangle has an area equal to the definite integral. Since f is continuous, its graph intersects the upper base of the rectangle.

At this stage, let us emphasize that **in applications, the quantities have dimensions that, in fact, are not the ones of an actual area.**

For instance, a physical power $P(t)$ integrated over time t from a to b is the physical work done during this time period. But it also corresponds to the area below the power function.

1.2.2 Fundamental Theorem of Integral Calculus

Let us consider the **indefinite integral with a variable upper limit**:

$$F(x) = \int_a^x f(u) \cdot du.$$

This means that the function in the variable u is integrated from the lower fixed limit a to the upper variable limit x.

Note: It is quite immaterial what the integration variable is called, as long as it is not the same as the integration limits.[3]

Theorem 8. *Let $f(x)$ be continuous and*
$$F(x) = \int_a^x f(u) \cdot du.$$

Then the derivative $F'(x)$ of $F(x)$ exists and $F'(x) = f(x)$.
In words: The derivative of an indefinite integral with respect to the upper limit equals the integrand evaluated at the upper limit.

Proof. We first compare the three areas on the interval $[x, x+h]$, where x is held **fixed**.

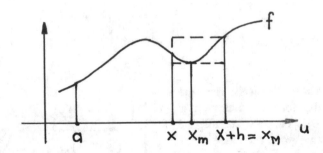

Let $f(x_M)$ be the global maximum and $f(x_m)$ the global minimum on the interval. Then the "sandwich relation" holds:

$$h \cdot f(x_m) \leq F(x+h) - F(x) \leq h \cdot f(x_M) \qquad \text{if } h > 0.$$

Dividing the two inequalities by h yields

$$f(x_m) \leq \frac{F(x+h) - F(x)}{h} \leq f(x_M).$$

Now we take the limit as $h \to 0$. As f is continuous, we have

$$f(x_m) \to f(x), \qquad f(x_M) \to f(x).$$

---

[3] Even in the literature one occasionally finds nonsensical $\int_a^x f(x)dx$.

The limit of the term in the middle of the sandwich is precisely the differential quotient $F'(x)$. This completes the proof of the Fundamental Theorem:

$$F' = f.$$

q.e.d.

It is remarkable that F is differentiable, as long as f is continuous. That is, the graph of F has a unique tangent at each of its points. For brevity, we say the graph of F is **smooth at each point**. Hence all possible corners (i.e. points of continuity at which the left-hand tangent differs from the right-hand tangent) of f are transformed into smooth points of F.

Distinct antiderivatives of the same function differ only by an additive constant. Therefore, when computing definite integrals, any **arbitrary antiderivative** F may be used:

$$\int_a^b f(x) \cdot \mathrm{d}x = F(b) - F(a).$$

The **set of all** antiderivatives of f is often denoted by

$$\int f(x) \cdot \mathrm{d}x \qquad \text{or more cautiously,} \qquad \int f(x) \cdot \mathrm{d}x + C.$$

Interesting and important for applications is the fact, that integral calculus may be extended to piecewise continuous functions f. Only at the jumps of f, the integral F has different left-hand and right-hand derivatives. At those points F is continuous but no longer uniquely differentiable: The graph of F has corners at those points.

Example 1.4 For each of the functions f there is a qualitative sketch of its indefinite integral $F(x) = \int_0^x f(u)\mathrm{d}u$:

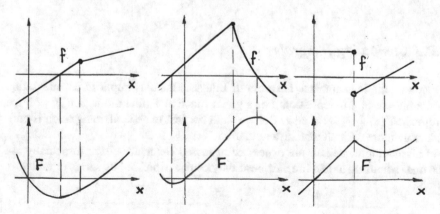

Summary:

- **Integration is a smoothing process:** Jumps are transformed into corners, corners are transformed into smooth points.
- **Differentiation is a roughing process:** Corners are transformed into jumps and it is possible that certain smooth points are transformed into corners.

Example 1.5

$$f(x) = \begin{cases} ax^2 & \text{if } x \geq 0 \\ 0 & \text{if } x < 0 \end{cases} \quad f'(x) = \begin{cases} 2ax & \text{if } x \geq 0 \\ 0 & \text{if } x < 0 \end{cases} \quad f''(x) = \begin{cases} 2a & \text{if } x \geq 0 \\ 0 & \text{if } x < 0 \end{cases}$$

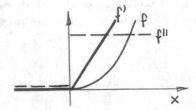

At $x = 0$, f is smooth, f' has a corner, and f'' has a jump discontinuity.
The functions f and f' are continuous everywhere, f'' is piecewise continuous (and discontinuous only at $x = 0$).   ◇

Example 1.6 Geometry of Railroad Tracks

When piecing together straight and curved sections of track one does not use circles, because the second derivative of

$$f(x) = \begin{cases} 0 & \text{if } x \leq 0 \\ r - \sqrt{r^2 - x^2} & \text{if } x > 0 \end{cases} \qquad \text{where r = radius,}$$

just as in the previous example of the parabola, is discontinuous at the point of transition (Verify!). When accelerated, this would result in a sudden jerk!   ◇

1.2.3 Computing Integrals

Evidently, the Fundamental Theorem of Integral Calculus implies that integration is the inverse of differentiation: For a given function f, find a function F with the property $F'(x) = f(x)$ for all x. But we shall see that in general, integration is considerably more difficult than differentiation.

Not all indefinite integrals are expressed in closed elementary form (especially not the ones occurring in real life). At best, they can be expressed by series or tables.

An example illustrating this is the integral of the so-called Gaussian distribution [4] $\int\limits_0^x e^{-u^2} du$, which plays a significant role in statistical applications.

Pocket calculators equipped with a symbolic calculator are capable of computing integrals that are closed expressions. They even handle parameters and are capable of making case distinctions. As a consequence, many integration techniques, used in earlier times for the tedious computation by hand, have now become obsolete. Therefore, when computing without calculator or computer, we restrict ourselves to integrals that can be dealt with by using the **Linear Property of Integral Calculus**

$$\int \{c \cdot f(x) + d \cdot g(x)\} \cdot dx = c \cdot \int f(x) \cdot dx + d \cdot \int g(x) \cdot dx$$

and the following properties A, B, and C:

A Basic integrals

$$\int x^r dx = \frac{1}{r+1} x^{r+1} + C, \; r \in \mathbf{R} \backslash \{-1\} \quad \int \frac{1}{x} dx = \ln|x| + C \quad \int e^{kx} dx = \frac{1}{k} e^{kx} + C$$

$$\int \sin(\omega t) dt = -\frac{1}{\omega} \cos(\omega t) + C \qquad \int \cos(\omega t) dt = \frac{1}{\omega} \sin(\omega t) + C$$

$$\int \sinh(ax) dx = \frac{1}{a} \cosh(ax) + C \qquad \int \cosh(ax) dt = \frac{1}{a} \sinh(ax) + C$$

B If the **integrand is the product of two functions, then integration by parts** holds:

$$\int f(x)g'(x) \cdot dx = f(x)g(x) - \int f'(x)g(x) \cdot dx + C.$$

Proof: By integrating the Product Rule $(fg)' = f'g + fg'$.

C If the **integrand contains the derivative of the inner function as a factor**, then

$$\int g(f(x)) \cdot f'(x) \cdot dx = G(f(x)) + C.$$

Here G is an arbitrary antiderivative of g. To prove this, integrate the Chain Rule:

$$[G(f(x))]' = g(f(x)) \cdot f'(x).$$

Often it is necessary to use a corrective constant factor, to be determined by differentiating a tentative antiderivative, see Example (c) below.

---

[4] Carl Friedrich Gauss (1777–1855) was a German mathematician and physicist. He was a universal genius who did outstanding scientific work in number theory, statistics, geodesy and astronomy. From 1816–1855, the Royal Observatory in Göttingen, which he headed, was his place of residence and work.

Example 1.7

(a) $\int x\sin(ax)dx = -\frac{1}{a}x\cos(ax) + \frac{1}{a}\int \cos(ax)dx = \frac{1}{a^2}[\sin(ax) - ax\cos(ax)] + C$

(b) $\int xe^{ax} \cdot dx = \frac{1}{a}xe^{ax} - \frac{1}{a}\int e^{ax} \cdot dx = \frac{1}{a^2}(ax - 1)e^{ax} + C$

(c) $\int \sqrt{ax^2 - x} \cdot (2ax - 1) \cdot dx = k \cdot (ax^2 - x)^{3/2} + C$

Due to $[(ax^2 - x)^{3/2}]' = \frac{3}{2} \cdot (ax^2 - x)^{1/2} \cdot (2ax - 1)$, we have $k = 2/3$.

(d) $\int (a+x)^n \cdot dx = \frac{1}{n+1}(a+x)^{n+1} + C$

(e) For $f(x) > 0$, we have $\int \frac{f'(x)}{f(x)} \cdot dx = \ln f(x) + C$ $\diamondsuit$

1.3 Applications of Integral and Differential Calculus

1.3.1 Escape Velocity

An object with mass m at a distance r from the earth's center ($r \geq$ earth's radius) and the earth with mass $M = 5.98 \cdot 10^{24}$ kg interact with each other due to the gravitational force

$$F(r) = G \cdot \frac{mM}{r^2}.$$

F is inversely proportional to the square of the distance, $G = 6.67 \cdot 10^{-11}$ m^{3}s^{-2}kg^{-1} is the universal gravitational constant.

On the earth's spherical surface with radius $R = 6,370$ km, the weight of the object is mg, where $g = \frac{GM}{R^2} \approx 9.81$ m/s^2 is the acceleration due to the earth's gravity.

The work W or energy needed to propel a body of mass m out of the gravitational field of the earth is

$$W = GmM \int_R^\infty \frac{dr}{r^2} = GmM\left[-\frac{1}{r}\right]_R^\infty = \frac{GmM}{R}.$$

The kinetic energy $\frac{m}{2}v^2$ for launching at speed v must at least be W:

$$\frac{m}{2}v^2 = \frac{GmM}{R} \implies v = \sqrt{\frac{2MG}{R}} \approx 11.2 \text{ km/s}.$$

The actual escape velocity must be larger, because we neglected the air drag.

1.3.2 Elasticity in Economics

In economics the elasticity is a meaningful tool to describe how the changes between the variables x and $y = f(x)$ relate to each other. Comparing the absolute changes Δf and Δx is often somewhat unsatisfactory. If, for instance, reducing the price of a product by 10 \$ implies an increase of the sales by 12,000 units, then one hardly has information about whether the reduction in price, compared to the price itself, is large or small. This deficiency is remedied by the elasticity function which compares the **relative** changes $\Delta x / x$ and $\Delta f / f$:

$$\frac{\Delta f / f}{\Delta x / x} = \frac{\Delta f}{\Delta x} \cdot \frac{x}{f}.$$

By taking the limit as $\Delta x \to 0$, we obtain the dimensionless **elasticity function of f with respect to** x,

$$\varepsilon_{f,x} = \frac{f'(x)}{f(x)} \cdot x. \tag{1.9}$$

Its value gives the percentage change in $f(x)$ in response to a 1% change in x.

Example 1.8 Let P be the price per unit if x units are sold. $P(x)$ is called the price function. If

$$P(x) = 1200 - x,$$

then

$$\varepsilon_{P,x} = \frac{-x}{1200 - x}.$$

For instance, if the demand is $x = 1000$, then $\varepsilon_{P,1000} = \frac{-1000}{200} = -5$. Hence if the quantity demanded at $x = 1000$ units increases by 1%, then the price reduces by 5%. If the quantity demanded at $x = 200$ units increases by 1%, then the price reduces by only 0.2%.   ◇

1.3.3 Harmonic Sum and Harmonic Series

We wish to use integrals to estimate the harmonic sum

$$H_n = 1 + \frac{1}{2} + \frac{1}{3} + \ldots + \frac{1}{n},$$

for which there is no closed form. To this end, we compare the areas in the following figure:[5]

---

[5] from [39].

On the one hand, H_n is the sum of rectangular areas and on the other hand, it is approximated by the area below the graph of $y = 1/x$.

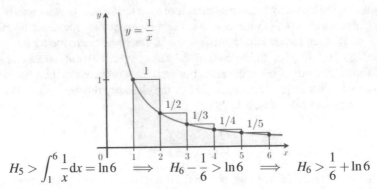

$$H_5 > \int_1^6 \frac{1}{x}\,dx = \ln 6 \implies H_6 - \frac{1}{6} > \ln 6 \implies H_6 > \frac{1}{6} + \ln 6$$

If we shift the six rectangles (including the rectangle with height $1/6$) one unit to the left, we obtain

$$H_6 - 1 < \ln 6 \implies H_6 < 1 + \ln 6.$$

In general, H_n satisfies the "sandwich relation"

$$\frac{1}{n} + \ln(n) \ < \ H_n \ < \ 1 + \ln(n). \tag{1.10}$$

Hence the harmonic series $\sum_{k=0}^{\infty} \frac{1}{k} = \infty$ diverges.

If we shift all the triangles that extend beyond the curve (and have a curved side downwards) to the left into the rectangle with $1 \le x \le 2$, then we immediately find

$$H_n - \ln n < 1.$$

The reason is that the sum of all n triangular areas is only a portion of the rectangular area equaling 1.

A more precise analysis concludes that

$$\lim_{n \to \infty} (H_n - \ln n) = \gamma,$$

where $\gamma \approx 0.5772156649$ is Euler's constant[6].

From the geometric point of view, it is evident that the total area of all the infinitely many triangles is slightly larger than half the largest rectangle on $1 \le x \le 2$, the latter having area 1.

---

[6] Leonhard Euler (1707–1783) was one of the most prominent mathematicians of all times. He did groundbreaking work in calculus, number theory, and physics.

1.3.4 Optimal Stopping

A tourist enjoying a cruise along the Rhine river knows that en route there will be 20 castles with overnight accommodations that will be accessed by the ship. She wishes to spend the night in the most beautiful castle. But she doesn't know at which cruise stop it will be, nor what it is called. At the sight of a castle she has to decide immediately whether or not to lodge there. There is no way back to a castle that has passed.

Question: What is the best strategy to maximize the probability of lodging in the most beautiful castle?

Let us consider the following strategy:
She lets the first $s - 1$ castles pass and denotes by B the most beautiful among them. Then she chooses among the remaining castles the first one that is more beautiful than B (if that never happens, then – bad luck – she has to lodge at the last castle).

The probability of finding the most beautiful castle A among the 20 castles depends on the choice of s. Let us denote this probability by $P(s)$. We now inquire into which s makes $P(s)$ a maximum.

- The probability that the most beautiful castle among the first $k - 1$ castles ($k > s$) is still B is the number of favorable outcomes divided by the number of possible outcomes, that is $(s-1)/(k-1)$.
- The probability that the kth castle is A equals $1/20$.

Consequently, the probability p_k that our tourist succeeds in choosing A as the kth castle is the product of the two probabilities, because the events are mutually independent:

$$p_k = \frac{1}{20} \cdot \frac{s-1}{k-1}.$$

Hence

$$P(s) = p_s + p_{s+1} + p_{s+2} + \ldots p_{20} = \frac{s-1}{20} \cdot \left(\frac{1}{s-1} + \frac{1}{s} + \frac{1}{s+1} + \frac{1}{s+2} + \ldots \frac{1}{19} \right).$$

More generally, if there are n castles, then in the equation above, 20 is replaced by n and 19 by $n - 1$.
For larger s and n, we may approximate the sum between the parentheses by the natural logarithm ln:

$$P(s) = \frac{s-1}{n} \cdot (H_{n-1} - H_{s-2}) \approx \frac{s-1}{n} \cdot [\ln(n-1) - \ln(s-2)] = \frac{s-1}{n} \cdot \ln \frac{n-1}{s-2}.$$

Furthermore, we may estimate

$$P(s) \approx \frac{s}{n} \cdot \ln \frac{n}{s}.$$

With $x = \frac{s}{n}$, we obtain the optimization function

$$f(x) = x \cdot \ln\frac{1}{x}.$$

Differentiation and equating to zero yields

$$\left(x \cdot \ln\frac{1}{x}\right)' = \ln\frac{1}{x} - x \cdot \frac{1}{x} = 0 \quad\Longrightarrow\quad \frac{1}{x} = e \quad\Longrightarrow\quad x_{\mathrm{opt}} = e^{-1}.$$

Whence $f(x_{\mathrm{opt}}) = f(e^{-1}) = e^{-1}$.

The graph of the function $f(x)$ confirms the result:

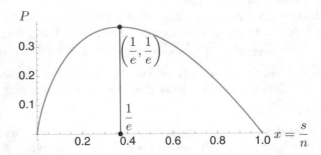

Interpretation and outlook:
This result is useful for $n > 9$ already.

Hence the optimal strategy is as follows: Wait until $1/e \approx 37\%$ of all castles have passed, then choose the first one that is more beautiful than the most beautiful among the first 37%. The probability of success is, after all, about 37%.

Of particular interest is the fact that the probability of success is practically independent of the size of n. That is, even in the case of 100 castles the probability is, quite surprisingly, still 37%. A plausible explanation is that with large n, also s becomes large, and hence the tourist has more information.

Another example utilizing the optimal stopping theory is the so-called Secretary Problem[7]. Here the issue is to find the most suitable among n applicants for a job opening, if the interviews are held one by one and the interviewer has to decide immediately whether to hire or to reject an applicant.
Remark: In real life the applications undergo a preselection of n people on the grounds of the documents they submitted.

---

[7] An investigation is found in [8].

1.3.5 Fermat's Principle, Snell's Law of Refraction

The Principle of Fermat[8] states that a ray of light in an optically isotropic medium (isotropic means that the physical properties are the same in all directions) follows a path from point P to point Q that requires the least time.

In an optical medium with refraction index $n \geq 1$ the speed of light is $c = c_0/n$, where $c_0 \approx 300,000$ km/h is the speed of light in a vacuum. Some values are given here:

air: $n = 1$ water: $n = 1.33$ glass: $1.5 \leq n \leq 1.62$

Let us analyze Fermat's Principle in the following case: The ray of light passes from the optically thin medium with (n_1, c_1) to an optically dense medium with (n_2, c_2); for instance, from air to water or glass. Then the above says that the speeds are inversely proportional to the refraction indices:

$$\frac{c_1}{c_2} = \frac{n_2}{n_1}.$$

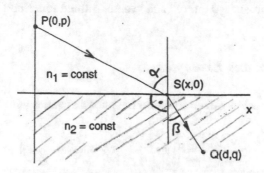

In the figure, the horizontal line is the interface between the two media (for example, air/water or air/glass). The positions of the points P and Q are given, S is a yet unknown point on the interface. Among all the bent light rays from P to Q after refraction at S, the one with shortest time is to be found. We have to solve an **optimization problem**:

The time required by a straight ray of light from P to S is the distance traveled/velocity, hence $T_1 = \frac{\sqrt{p^2+x^2}}{c_1}$.   From S to Q it is $T_2 = \frac{\sqrt{(d-x)^2+q^2}}{c_2}$.

Hence the total time from P to Q is

$$T(x) = \frac{\sqrt{p^2+x^2}}{c_1} + \frac{\sqrt{(d-x)^2+q^2}}{c_2}.$$

---

[8] Pierre Fermat (1601-1665) was a full-time lawyer. But he did seminal work in number theory, which was his passion. His famous conjecture that none of the integer equations $a^n + b^n = c^n$ with $n \geq 3$ has integer solutions has held many generations of mathematicians busy. Andrew Wiles succeeded in proving the conjecture of the century during the years around 1993. The book [34] tells this fascinating story brilliantly.

The derivative with respect to x is zero (use the Chain Rule):

$$\frac{dT}{dx} = \frac{x}{c_1 \cdot \sqrt{p^2 + x^2}} - \frac{d - x}{c_2 \cdot \sqrt{(d - x)^2 + q^2}} = 0.$$

Expressed in terms of the two angles, this gives rise to **Snell's Law of Refraction**:

$$\frac{\sin \alpha}{c_1} - \frac{\sin \beta}{c_2} = 0 \quad \Longrightarrow \quad \frac{\sin \alpha}{\sin \beta} = \frac{c_1}{c_2} = \frac{n_2}{n_1}.$$

In a dense medium the light ray is bent towards the normal of the interface.

1.4 Parametrized Curves or Vector-Valued Functions

The notions **"parametrized curve"** and **"vector-valued function"** are equivalent.

1.4.1 Definition and Examples

Let us consider a **vector-valued function in space** that varies with the parameter t:

$$\vec{r}(t) = \begin{pmatrix} x(t) \\ y(t) \\ z(t) \end{pmatrix}, \quad t \in \text{interval}$$

Here $\vec{r}(t)$ is the position vector depending on t with Cartesian coordinates $x(t), y(t), z(t)$ that depend on t.

Of course, the symbol for the parameter can be chosen arbitrarily as is common for functions. If it has the physical meaning of time, then it is customary to use t. In this case the vector-valued function may describe the position of an object in space (for example, a satellite) at any time t.

The notion "vector-valued function" is justified by the fact that $\vec{r}(t)$ can be interpreted as a function $t \mapsto \vec{r}(t)$.

The domain is an interval, the range consists of position vectors.

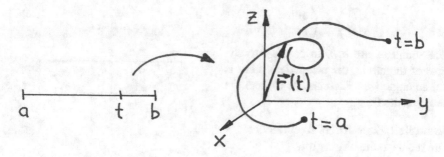

Two remarks:

- If t is the time, then the curve described by $\vec{r}(t)$ does **not only contain the geometry of the path, also called the orbit, but also the motion in the course of time.**

- Often a curve can only be described by a parametric equation, for instance, if it is a space curve. But plane curves, too, may at times be more easily described by parametric equations. The parametric equation is more general and more flexible than, say, descriptions of the form $y = f(x)$. To illustrate this, let us consider the following examples, with t being the time.

Example 1.9 Straight Line

The figure shows a straight line in space passing through A with direction vector $\vec{b}$. Let $\overrightarrow{OA} = \vec{a}$.

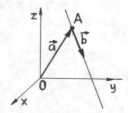

$$\vec{r}(t) = \vec{a} + t \cdot \vec{b} = \begin{pmatrix} x(t) \\ y(t) \\ z(t) \end{pmatrix} = \begin{pmatrix} a_1 + tb_1 \\ a_2 + tb_2 \\ a_3 + tb_3 \end{pmatrix}.$$

The vector $\vec{b}$ has the meaning of the motion's constant velocity.   ◇

Example 1.10 Circle

The figure shows a circle with radius R.

$$\vec{r}(t) = \begin{pmatrix} x(t) \\ y(t) \end{pmatrix} = R \cdot \begin{pmatrix} \cos t \\ \sin t \end{pmatrix}.$$

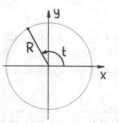

Here the parameter has the additional geometric meaning of an angle.

Example 1.11 Ellipse

$$\vec{r}(t) = \begin{pmatrix} x(t) \\ y(t) \end{pmatrix} = \begin{pmatrix} a\cos t \\ b\sin t \end{pmatrix}.$$

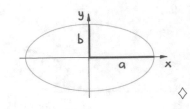

The semiaxes are a, b. In contrast to the case of the circle, the parameter t here is not an angle (see Exercise 31: Concentric circle method). ◇

Example 1.12 Spiral of Archimedes

The radius to the origin grows linearly with the angle t:

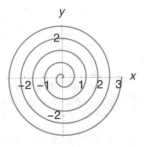

$$\vec{r}(t) = at \begin{pmatrix} \cos t \\ \sin t \end{pmatrix}.$$

◇

Example 1.13 Logarithmic Spiral

$$\vec{r}(t) = e^{at} \begin{pmatrix} \cos t \\ \sin t \end{pmatrix}.$$

The radius to the origin grows exponentially with the angle t.

The following Mathematica command produces the left-hand figure and can also be entered on the Internet in the free application **Wolfram Alpha**:

```
ParametricPlot[Exp[-0.05 t]{Cos[t],Sin[t]},{t.0,50}]
```

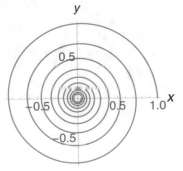

In nature we often encounter logarithmic growth. A particularly graceful example is the shell of the nautilus[9]. When sliced open, the chambers show a discrete logarithmic dependence in their sizes. ◇

---

[9] with kind permission of Pixabay: This photograph is not subject to copyright.
Link https://pixabay.com

Example 1.14 Cycloid
This is the curve traced out by a point P on the circumference of a wheel as it rolls along a straight line. Here t is the rolling angle.
We obtain its parametrization by inspecting the following figure:

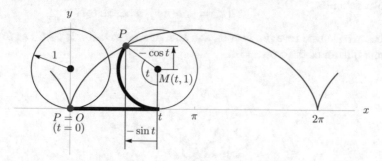

$$\overrightarrow{OP} = \overrightarrow{OM} + \overrightarrow{MP} = a\begin{pmatrix} t \\ 1 \end{pmatrix} + a\begin{pmatrix} -\sin t \\ -\cos t \end{pmatrix}. \tag{1.11}$$

Summarized,

$$\vec{r}(t) = a\begin{pmatrix} t - \sin t \\ 1 - \cos t \end{pmatrix}. \tag{1.12}$$

In Example 1.18 we shall show that the cycloid has cusps i.e. corners with vertical tangents at the points where $y = 0$. ◇

Example 1.15 Helix
The helix is given by the radius r and the pitch h per revolution, where the pitch increases linearly with the angle t:
$$\vec{r}(t) = \begin{pmatrix} r\cos t \\ r\sin t \\ \frac{h}{2\pi}t \end{pmatrix}.$$

The Mathematica command
`ParametricPlot3D[{Cos[t] ,Sin[t], 0.05t},{t.0,8Pi}]`
produces the following figure:

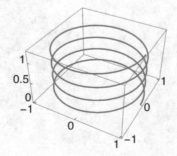

◇

Plane curves may also be described by **polar equations**. Here the distance $r(\varphi) \geq 0$ is a function of the angle φ. Command: `PolarPlot[...]`.

Example 1.16 Maurer Roses

The polar equation

$$R(\varphi) = |\sin(n\varphi)| \text{ where } n \in \mathbf{N} \tag{1.13}$$

describes a $2n$-leaved rose if n is even, and an n-leaved rose if n is odd. The following figure illustrates the case $n = 4$:

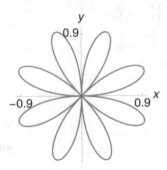

Let us consider 360 discrete points on the curve (1.13) corresponding to the angles $\varphi = 0°, 1°, \ldots, 359°$. Peter M. Maurer discovered that beautiful diagrams are created, if consecutive points corresponding to angles that differ by $d°$ are joined linearly.[10]

If we choose a number d that is not a divisor of 360, then all 360 points are accessed. If we choose d to be a divisor of 360, then only a partial set of points are accessed. In the latter case we repeat the process of joining points with a starting point that has not yet been used. This process is repeated until all points have been accessed.

Here are two examples from [46]:

$n = 5, d = 120$ $\qquad\qquad\qquad\qquad\qquad$ $n = 6, d = 72$

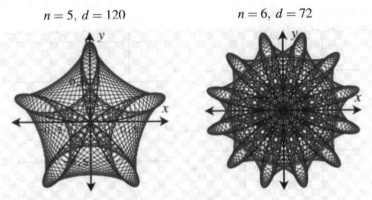

More figures showing the beauty of geometry are found in [29].

---

[10] See publication [26].

1.4.2 Ellipse

Definition 9. *The curve described by the Cartesian equation*

$$\frac{x^2}{a^2} + \frac{y^2}{b^2} = 1 \qquad where \quad a \geq b > 0$$

is called an ellipse with semiaxes a and b.

Because $y = \pm\frac{b}{a}\sqrt{a^2 - x^2}$, the ellipse is an affine circle. By that we mean a circle that is compressed in the direction of the y-axis by a factor $\frac{b}{a} \leq 1$. The radical expression describes a circle with radius a.

According to the following figure, the foci F_1, F_2 of the ellipse are defined as the points whose distances to the center C is $e = CF_1 = CF_2 = \sqrt{a^2 - b^2}$.

The quantity e is called the **linear eccentricity** of the ellipse. The physical notion of the focus is based on the fact that every ray of light from one focus is reflected from the ellipse (angle of incidence = angle of reflection) to the other focus.

A circle of radius r is a special case of an ellipse: $a = b = r$ (both semiaxes are equal).

With the aid of the following figure we deduce the polar equation of the ellipse, which will be useful when solving the two-body problem in Section 4.13.

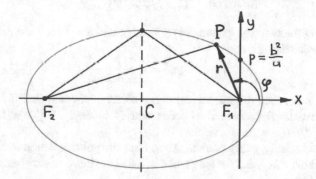

Desired is $r(\varphi)$. The equation of an ellipse centered at $C(-e, 0)$ is

$$\frac{(x+e)^2}{a^2} + \frac{y^2}{b^2} = 1.$$

Substituting the polar coordinates $x = r\cos\varphi, y = r\sin\varphi$ yields

$$\frac{(r\cos\varphi + e)^2}{a^2} + \frac{(r\sin\varphi)^2}{b^2} = 1.$$

Multiplying both sides by the denominators implies

$$b^2(r^2\cos^2\varphi + 2er\cos\varphi + e^2) + a^2(r^2\sin^2\varphi) = a^2b^2.$$

Collecting like powers of r yields the quadratic equation

$$r^2(b^2\cos^2\varphi + a^2\sin^2\varphi) + r(2eb^2\cos\varphi) + (e^2b^2 - a^2b^2) = 0.$$

The expression inside the last set of parentheses equals $(a^2 - b^2)b^2 - a^2b^2 = -b^4$.
The expression inside the first set of parentheses equals
$(a^2 - e^2)\cos^2\varphi + a^2\sin^2\varphi = a^2 - e^2\cos^2\varphi$.
Norming the quadratic equation for r yields

$$r^2 + 2\frac{eb^2\cos\varphi}{a^2 - e^2\cos^2\varphi}r - \frac{b^4}{a^2 - e^2\cos^2\varphi} = 0.$$

With $N = a^2 - e^2\cos^2\varphi$ the solutions are

$$r(\varphi) = -\frac{eb^2\cos\varphi}{N} \pm \sqrt{\frac{e^2b^4\cos^2\varphi}{N^2} + \frac{b^4}{N}}.$$

As $r > 0$, only the $+$ sign holds. Simplifying the radical expression

$$\frac{1}{N}\sqrt{e^2b^4\cos^2\varphi + b^4(a^2 - e^2\cos^2\varphi)} = \frac{ab^2}{N}$$

yields

$$r(\varphi) = \frac{1}{N}(-eb^2\cos\varphi + ab^2) = \frac{b^2(a - e\cos\varphi)}{(a + e\cos\varphi)(a - e\cos\varphi)}$$

$$= \frac{b^2}{(a + e\cos\varphi)} = \frac{b^2/a}{(1 + e/a\cos\varphi)}.$$

By introducing the two new parameters $p = b^2/a$ and $\varepsilon = e/a$, we obtain the final
form of the **polar equation**:

$$r(\varphi) = \frac{p}{1 + \varepsilon\cos\varphi}.$$

The quantity ε is called the **numerical eccentricity** of the ellipse. Similar ellipses
have, therefore, the same numerical eccentricity. Obviously, $0 \le \varepsilon < 1$, while the
limiting case $\varepsilon = 0$ describes a circle.
Remark: Interestingly, every conical section has the polar equation derived above.
For the hyperbola $\varepsilon > 1$, and for the parabola, being the limiting case between ellipse
and hyperbola, $\varepsilon = 1$.
Finally, let us check the polar equation at distinguished points:

$$r(0) + r(\pi) = \frac{p}{1 + \varepsilon} + \frac{p}{1 - \varepsilon} = \frac{2p}{1 - \varepsilon^2} = \frac{2b^2/a}{1 - (a^2 - b^2)/a^2} = \frac{2b^2/a}{b^2/a^2} = 2a.$$

1.4.3 Tangents of a Curve, Vectors of Velocity

Let the motion of an object in space or in the plane be described by the time-varying
position vector $\vec{r}(t)$. Let us compute the **velocity** $\vec{v}(t)$ **and the speed** $v = |\vec{v}|$. The
velocity is a vector quantity. The speed is its magnitude, hence a scalar quantity.

We now consider a small increment of time Δt during which the object moves from P to Q. Then the average velocity on the straight path is

$$\vec{v}_m = \frac{\Delta \vec{r}}{\Delta t} = \frac{\vec{r}(t + \Delta t) - \vec{r}(t)}{\Delta t}.$$

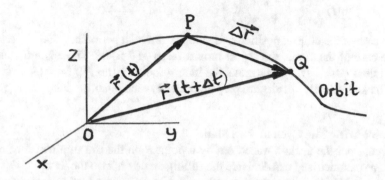

By letting $\Delta t \to 0$ we obtain the **instantaneous velocity** $\vec{v}(t)$ at time t.

The velocity vector $\vec{v}(t)$ is obviously **tangent** to the orbit at point $\vec{r}(t)$.
Performing the calculations yields

$$\vec{v}(t) = \lim_{\Delta t \to 0} \frac{1}{\Delta t} \begin{pmatrix} x(t + \Delta t) - x(t) \\ y(t + \Delta t) - y(t) \\ z(t + \Delta t) - z(t) \end{pmatrix} = \lim_{\Delta t \to 0} \begin{pmatrix} \frac{x(t+\Delta t)-x(t)}{\Delta t} \\ \frac{y(t+\Delta t)-y(t)}{\Delta t} \\ \frac{z(t+\Delta t)-z(t)}{\Delta t} \end{pmatrix} = \begin{pmatrix} \dot{x}(t) \\ \dot{y}(t) \\ \dot{z}(t) \end{pmatrix}.$$

Hence the velocity vector is obtained by coordinatewise differentiation of the position vector.

It is customary to denote derivatives with respect to time by a dot rather than by a prime. In short notation,

$$\vec{v}(t) = \dot{\vec{r}}(t).$$

The speed is the magnitude of the velocity: $v(t) = \sqrt{\dot{x}(t)^2 + \dot{y}(t)^2 + \dot{z}(t)^2}$.

More generally, even if the parameter does not have the meaning of time, **the derivative of the position vector is always tangent** to the curve!

Of course, if the problem is planar, the z-coordinate disappears.

Example 1.17
Let us consider the following two motions of an object at times $t \geq 0$:

$$\vec{r}_1(t) = \begin{pmatrix} a\cos t \\ b\sin t \end{pmatrix} \qquad \vec{r}_2(t) = \begin{pmatrix} \cos(t^2) \\ \sin(t^2) \end{pmatrix}.$$

The respective orbits are an ellipse and a circle.

The velocities and speeds are

$$\vec{v}_1(t) = \begin{pmatrix} -a\sin t \\ +b\cos t \end{pmatrix}, \qquad |\vec{v}_1(t)| = \sqrt{a^2\sin^2 t + b^2\cos^2 t},$$

$$\vec{v}_2(t) = 2t \cdot \begin{pmatrix} -\sin(t^2) \\ +\cos(t^2) \end{pmatrix}, \qquad |\vec{v}_2(t)| = 2t.$$

In the case of the ellipse, position and velocity are both periodic.

In the case of the circle, the object moves faster and faster. The angle traveled in the circle is $\varphi(t) = t^2$. Furthermore, the inner product $\vec{r}_2(t) \cdot \vec{v}_2(t) = 0$ confirms that position and velocity vectors are perpendicular to each other.

$$\diamond$$

Example 1.18 The Cycloid Revisited.

We reconsider the cycloid traced out by a point P on the rim of a wheel of radius $a = 1$, as parametrized earlier. Here the rolling angle t is also the time:

$$\vec{r}(t) = \begin{pmatrix} t - \sin t \\ 1 - \cos t \end{pmatrix} = \begin{pmatrix} t \\ 1 \end{pmatrix} + \begin{pmatrix} -\sin t \\ -\cos t \end{pmatrix}.$$

The wheel's center evidently moves to the right at a constant speed. Coordinatewise differentiation yields the velocity of point P as a function of time:

$$\vec{v}(t) = \dot{\vec{r}}(t) = \begin{pmatrix} 1 - \cos t \\ \sin t \end{pmatrix}.$$

The speed is

$$v(t) = \sqrt{(1 - \cos t)^2 + \sin^2 t} = \sqrt{2(1 - \cos t)}.$$

Obviously $v(0) = 0$. The corner of the cycloid is an instantaneous rotational center. The maximum speed is $v(\pi) = 2$, i.e. twice that of the wheel's center.

As we check the answer it becomes evident that at maximum speed the velocity $\vec{v}(\pi) = \begin{pmatrix} 2 \\ 0 \end{pmatrix}$ is directed horizontally.

Next let us show by computation that the cycloid has so-called **cusps**, that is, corners with vertical tangents:

The velocity vectors are tangents to the curve with angle of inclination α given by

$$\tan \alpha = \frac{\dot{y}(t)}{\dot{x}(t)} = \frac{\sin t}{1 - \cos t}.$$

But for $t = 0$, the velocity vector disappears, so we have to take the limit as $t \to 0$. Moreover, the Rule of Bernoulli-L'Hôpital applied to the above yields $\lim_{t \to 0} \dfrac{\cos t}{\sin t} = \infty$, implying a vertical tangent.

If we consider the velocity vector in the form

$$\vec{v}(t) = \dot{\vec{r}}(t) = \begin{pmatrix} 1 \\ 0 \end{pmatrix} + \begin{pmatrix} -\cos t \\ \sin t \end{pmatrix},$$

then it becomes clear that the motion is the superposition of a translation to the right and a rotation about the wheel's center.

Visualization (The diagram on the right shows various resulting velocities):

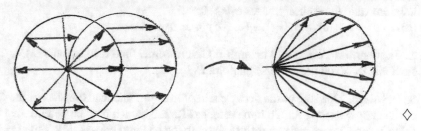

Example 1.19 Self-Similarity of the Logarithmic Spiral
The logarithmic spiral $r = e^{a\varphi}$ multiplied by a factor k is mapped onto itself by a rotation through some angle α. The reason is that

$$k \cdot e^{a\varphi} = e^{\ln k + a\varphi} = e^{a(\varphi + \frac{\ln k}{a})} \qquad \text{where } \alpha = \frac{\ln k}{a}.$$

In addition, the angle between the position vector and the tangent vector is constant, because stretching preserves angles.

◊

1.5 Exercises Chapter 1

1. Exponential Functions. By changing the bases to e, prove that the following laws are valid for arbitrary bases $a, b > 0$:
(a) $a^x \cdot b^x = ?$ (b) $a^x \cdot a^y = ?$ (c) $(b^x)' = \ln b \cdot b^x$.

2. Exponential Function Through a Given Point. Find a function $f(x) = Ce^{0.4x}$ such that its graph passes through point $P(1, 1/2)$.

3. Uranium. One step of the decay chain of uranium into lead consists of the decay of the radium isotope Ra226 into radon gas Rn222, which may be potentially harmful to human beings and can diffuse through the soil into houses. If at time $t = 0$ the amount of radium is m_0, the amount of radon gas after t years is approximately

$$m(t) = m_0 \frac{\lambda_1}{\lambda_2 - \lambda_1} \{e^{-\lambda_1 t} - e^{-\lambda_2 t}\}, \text{ where the annual decay constant is } \lambda_1 = \ln 2/1602$$

for Ra226 and $\lambda_2 = \ln 2/0.01048$ for Rn222.
(a) What are the half-lives of radium and radon?
(b) Determine the amount of radon at $t = 0$ and $t \to \infty$.
(c) When does radon reach its maximum amount?

4. Brainteaser: Sprinter and Tortoise. A sprinter and a tortoise race against each other. Their respective constant speeds are assumed to be 10 m/s and 1 m/s.
Claim: If the tortoise has a head start of 10 m, then the sprinter will never pass the tortoise!
Proof: If the sprinter runs 10 m, then the tortoise is 1 m ahead, if the sprinter runs another 1 meter, then the tortoise is 10 cm farther ahead, if the sprinter runs another 10 cm, then it is 1 cm ahead, ... This process goes on indefintitely.
What is the fallacy in the above argument? When and where will the sprinter overtake the tortoise?

5. Gaussian Bell Curve. The Gaussian bell curve is important in statistics. It is given by

$$y = e^{-kx^2} \text{ where } k > 0.$$

Discuss the function, compute the coordinates of the inflection points, and sketch the graph.

6. Hyperbolic Functions. Show that

$$\cosh(2x) \text{ and } \sinh(2x)$$

satisfy identities similar to their trigonometric equivalents.

7. Logarithms.
(a) Compute analytically and numerically the conversion factor a for transforming $\ln x$ into $\log_{10} x = a \cdot \ln x$ and compare the two graphs.
(b) Derive the two laws of logarithms $\log_b(u \cdot v) = ?$ and $\log_b(u^r) = ?$.

8. Growth Model for Young Children. An empirical equation for the statistical average height h in cm as a function of the age t in years for the time period $\frac{1}{4} \leq t \leq 6$ is given by

$$h(t) = 70.23 + 5.104 \cdot t + 9.22 \cdot \ln t.$$

(a) Compute the height and the growth rate in cm/year of a two-year old child.

(b) When does the growth rate in cm/year reach its maximum?

9. Quotient Rule. Let f and g be two given functions. Deduce the Quotient Rule for differentiation with the aid of the Product and Chain Rules:

$$\left[\frac{f}{g}\right]' = \frac{f'g - fg'}{g^2}.$$

10. Least Square Method. Let $x_1, x_2, \ldots, x_n$ be the measured values of a physical quantity. Measurements are known to be deficient. Compute from the measured data a value x such that the sum of the squares of the deviations $(x - x_i)^2$ is a minimum. Give an interpretation of your answer.

11. Puzzling Worm's Eye View. We assume that the earth's surface with radius $R = 6370$ km is a perfectly smooth spherical surface.

How far can a human being or an animal view, if the eyes are at an altitude h above the earth's surface?

Hint: Assume that the eyes are at point P above a circle with giant radius R. Let D be the lengths of the two tangent segments from P to the circle. D is the maximum viewing distance. Use the Tangent-Secant Theorem $D^2 = h(2R + h)$ and compute numerically $D(h)$ for various orders of magnitudes of h.

12. Maximum Consumer Power. A battery-driven device is to be built. The battery with voltage U has a given internal resistance R_i. What should the resistance R of the load be, the load being the consumer, in order to drain a maximum consumer power P from the device?

Physical facts: The two resistors in the circuit are connected in series. Hence the current is $I = \frac{U}{R_i + R}$ and the consumer power is

$$P(R) = U \cdot I = R \cdot I^2 = \frac{U^2 R}{(R_i + R)^2}.$$

13. Optimal Packaging. Determine the dimensions of a cylindrical tin can with volume 1 liter that minimizes the material. Assume that bottom, top and lateral surface are manufactured from the same sheet metal.

What is the ratio between diameter d and height h?

Compare with real-life cans where diameter : height = 1 : 1.18. Explain.

14. Salmon Swimming with Minimum Energy. In order to reach their spawning ground s km upstream, the salmon swim against the current with a velocity r relative to the water. The river has a constant velocity v. The swimming power P is determined by the friction of water to be overcome. Biologists found by measurements that P is proportional to r^λ with some constant $\lambda > 2$.

What relative velocity $r > v$ minimizes the energy W spent during the journey to the spawning ground?

15. Elasticity of a Function. Let the demand function $N(p) = (p-20)^2$ be given as a function of the price p per unit.

(a) Sketch the graph of the function $N(p)$.

(b) Compute the elasticity function $\varepsilon_{N,p} = \frac{N'(p)}{N(p)} \cdot p$.

(c) Compute the elasticity at $p = 5$ and $p = 19$ and give an interpretation of your answers.

16. Integration. Compute by hand and check your answers by differentiating:

(a) $\int (ax^3 + b\sqrt{x}) \cdot dx$ (b) $\int x\cos(kx) \cdot dx$

(c) $\int xe^{(x^2)} \cdot dx$ (d) $\int \frac{\cos x}{1 + \sin x} \cdot dx$

17. Brainteaser: Snowplow. On a certain day it snows with constant intensity. A snowplow starts work at 8 a.m. At 9 a.m. it has covered 2 km, at 10 a.m. it has covered 3 km. When did it start snowing (provided the snowplow always removes the same amount of snow per unit time)?

18. Brainteaser: Spider Kunigunde.[11] An elastic thread of length $1,000$ km is firmly clamped at point A, see figure. Kunigunde starts crawling on the thread at A and moves forward with a constant speed of $v = 1\,\mathrm{mm/s}$. After each full second the thread is extended abruptly (i.e. there is no time lapse) by pulling the free end E through another $1,000$ km.

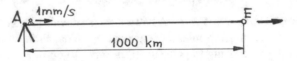

(a) Prove that Kunigunde does reach the end E of the thread.

(b) Determine her age when she arrives at E.

19. Tedious Integrals. Convince yourself by use of a pocket calculator or a computer that

$$\int \frac{\sin x}{x} \cdot dx$$

is a nonelementary integral. Find more such seemingly simple integrals by testing.

20. Kepler's Barrel Rule (Newton-Cotes formula). Show that for every polynomial of $\leq$ 3rd degree

$$q(x) = ax^3 + bx^2 + cx + d,$$

this surprising identity holds:

$$\int_0^h q(x)dx = h \cdot \left[\frac{1}{6}q(0) + \frac{2}{3}q(h/2) + \frac{1}{6}q(h) \right].$$

---

[11] This brainteaser is based on an oral communication by my friend Dr. ing. ETH Daniel Rufer, who in 2016 lost his life due to a bicycle accident. This problem is dedicated to him posthumously.

That is, except for the factor h, the integral is the weighted average of the three function values at $x = 0$, $\frac{h}{2}$, h with weights $\frac{1}{6}$, $\frac{2}{3}$, $\frac{1}{6}$.

Historical facts:

Kepler, when buying wine in a barrel of height h, was suspicious when the wine merchant determined the volume of the barrel by just one single length measurement with a measuring rod through a taphole. Thereupon he derived the formula above to obtain a more accurate approximation of the volume by three measurements at the bottom, the top and in the middle. Here the three q-values correspond to the three circular cross-sectional areas.

21. Checking Integrals. Pocket calculators and computers provide the following answer:

$$\int e^{-kt} \cos(\omega t)\,dt = \frac{1}{k^2 + \omega^2} e^{-kt} \{\omega \sin(\omega t) - k\cos(\omega t)\} + C.$$

Verify the correctness by differentiating by hand or by use of a computer, then plot the integrands for $k = 0.1$ and $k = 0.01$.

22. Geometry of Railway Tracks, Roads and Ski Jumps.
We wish to find a suitable transition curve between two straight line segments, two circular arcs, or between a straight line segment and a circular arc.

(a) Show that a cubic connection to a straight segment has a continuous second derivative at the point of transition. A vehicle could therefore pass the transition point without experiencing a sudden jerk.

(b) In road or track design one commonly uses the so-called clothoid[12], which is characterized by its curvature $K = 1/R$ (R is the radius of the circle of tangency) being proportional to its distance ℓ along the track. The parametric equation in Cartesian coordinates is

$$x(\ell) = A \cdot \sqrt{\pi} \int_0^\ell \cos(\frac{\pi}{2}u^2)\,du$$

$$y(\ell) = A \cdot \sqrt{\pi} \int_0^\ell \sin(\frac{\pi}{2}u^2)\,du$$

Clothoids are all similar to each other via the factor A.
Here is the graph for $A = 1$:

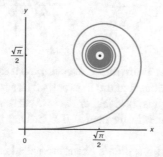

---

[12] I am obliged to dipl. math. P. Fässler for referring me to the clothoid.

(i) Verify that the curvature

$$K = \frac{\dot{x}\ddot{y} - \ddot{x}\dot{y}}{(\dot{x}^2 + \dot{y}^2)^{\frac{3}{2}}} \qquad \text{where} \qquad \dot{} = \frac{d}{d\ell}$$

varies linearly with ℓ.

(ii) Euler was the first to compute this curve in 1743. Only in 1781 he succeeded in proving that the improper integrals satisfy

$$\int_0^\infty \cos(\frac{\pi}{2}u^2)du = \int_0^\infty \sin(\frac{\pi}{2}u^2)du = \frac{1}{2}.$$

Verify that this implies that the curve converges to the limit point $(A\frac{\sqrt{\pi}}{2}, A\frac{\sqrt{\pi}}{2})$ and that $\ell, K \in [0, \infty)$. This means that a transition curve with any desired curvatures at the two ends may be constructed.

(iii) Compute the speed $v(\ell)$ of a vehicle running along a clothoid, if its parameter ℓ has the meaning of time. Also compute $L(\ell) \cdot R(\ell)$, where $L(\ell)$ is the length of the curve between the points $(0,0)$ and $(x(\ell), y(\ell))$.

(iv) According to [41], the run-up to a ski jump consists of 3 sections:

An inclined line segment, followed by an arc of a clothoid with increasing curvature (whereby the initial curvature is of course 0 in order to achieve a smooth transition) and finally the take-off table, a short line segment that is inclined slightly downwards, tangential to the end point of the curve.

The transition from the clothoidal arc to the take-off table results in an abrupt decrease in the maximum centrifugal force to 0. This effect is welcome and facilitates the jumper's take-off.

The maximum centrifugal force should not exceed 70% of the jumper's weight. Let the jump speed be 100 km/h on a large hill. What is the final radius of the clothoidal arc?

23. Piecewise Linear Functions. Compute and sketch

(a) $G(x) = \int\limits_{-1}^{x} g(u)du$ for the displayed function g.

(b) f' for the displayed function f.

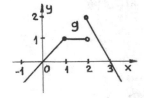
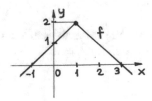

24. Brainteaser: Express Train and Continuity. An express train passes a 200 km section between the two cities A and B in exactly 2 hours. It has to decelerate at times because of construction sites, bridges, or narrow curves. But on straight sections, it can temporarily travel much faster than 100 kph. That is, the velocity changes with time.

Show that, irrespective of the train's changing speeds, there is always at least one precisely 100 km-portion of the section that can be traveled in exactly 1 hour's time.

25. Concentric Circle Method for an Ellipse and Parametrizing a Hyperbola.

(a) Show that points P, as drawn in the figure, lie
on an ellipse with semiaxes a, b. By inspecting
the figure verify that the parameter t is not the
angle corresponding to the point P.

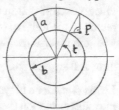

(b) Show that the parametrized curve

$$\begin{pmatrix} x(t) \\ y(t) \end{pmatrix} = \begin{pmatrix} a\cosh t \\ b\sinh t \end{pmatrix}$$

   satisfies the equation $\qquad \dfrac{x^2}{a^2} - \dfrac{y^2}{b^2} = 1$

and thus describes a hyperbola.

26. Lissajous Figure. Let the following vector-valued function be given:

$$\vec{r}(t) = \begin{pmatrix} \sin(2t) \\ \sin t \end{pmatrix} \qquad \text{where} \quad 0 \le t \le 2\pi.$$

(a) Sketch the graph, compute and mark some special points together with their
   t-values.
(b) Compute the angle at the point of intersection.
(c) Check whether the tangent line at the highest point is horizontal.

27. Controlling a Milling Cutter.

A milling cutter with radius R is to cut out an
arc of the parabola $y = ax^2$, see figure. Find
the parametrized orbit of the center $\vec{r}_F$ of the
milling cutter.
Hint: First parametrize the parabola.

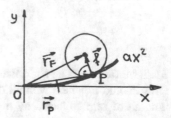

28. Self-Similarity of the Logarithmic Spiral Revisited. By use of the inner product, confirm the earlier finding that the angle between the position vector and the
tangent vector is constant.

29. Pedal Curve

(a) Curtate and Prolate Cycloids

   (i) Compute the parametric equation of the curve traced out by a point at a distance r from the center of a wheel of radius 1 rolling on a straight horizontal
   line, where the rolling angle = time = t.
   Applications: A luminous body attached to a spoke describes a curtate cycloid
   ($r < 1$), a wavy curve like a serpentine. A point on the wheel-flange of a
   railroad locomotive traces out a prolate cycloid ($r > 1$), a curve containing
   loops.

(ii) Compute the velocities at the highest and the lowest points of the curve. Convince yourself that in the case of $r > 1$ the velocity at the lowest point is directed to the left, i.e. opposite to the wheel's direction, which is in accordance with the reverse motion in the loop.

(b) Path of a Bicycle Pedal

(i) Let us consider a bicycle with wheels of radius 1 and a pedal at a distance $\frac{1}{2}$ of the axis of rotation (which is at the same height as the wheels' axles). Compute the parametric equation of the pedal's path as a function of the rolling angle = time = t. Take into account a given gear ratio. Use the lowest point of the pedal as the origin of the reference frame. Denote by ωt the rotation angle of the crank. Then the quantity ω describes the gear.

(ii) Compute the velocities at the highest and the lowest points. For which ω are there loops, like in the case of a prolate cycloid? For which ω are there no loops, like in the case of a curtate cycloid?

(iii) Show that for a certain value of ω, the pedal curve is affine to a cycloid with rolling radius $\frac{1}{2}$, and hence has cusps.

30. Wheel on a Circle.

Parametrize the curve (see figure) traced out by a point on the circumference of a wheel with radius $r = 1$ rolling along a fixed circle with radius $R = 3$. This is called an epicycloid.

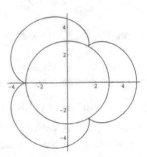

31. Brainteaser: The Rain Problem.

How should a person without umbrella walk in the rain in order to minimize the amount of rain collected by his body? It's a popular topic and is discussed controversially. For simplicity, assume that the person's front and back are flat with a certain area (imagine a sheet metal with the contours of a human with head, body, and legs) and assume that the person moves upright along a straight line segment of length s on the plane at constant speed v. In addition, assume that the rain has constant speed r and intensity everywhere.

(a) Determine the optimal speed v for the two special cases in which the rain pelts the person horizontally from the front or from behind.

(b) Do the same for any direction of rain speed from the front or the rear.

(c) Explain that the result in (b) is still valid for a human body if we choose for the plane surface the projection of the body perpendicular to the direction of travel. However, what has been neglected?

32. Fountain. A fountain having rotational symmetry consists of a large number of small water jets. The figure shows a vertical cross-section through the axis of symmetry. We observe that the jets ejected from the center have varying inclination angles α. Assume that all water particles emitted from the fountain have the same initial velocity v_0 and that the friction of air is negligible.

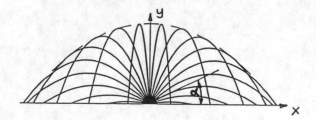

(a) Describe the paths of the water particles in the form $\vec{r}(t) = \begin{pmatrix} x(t) \\ y(t) \end{pmatrix}$ depending on the parameter α, while t is the time.

(b) Describe the paths in the form $y = f(x)$ by eliminating t.

(c) Determine the coordinates (x_m, y_m) of the highest points of the paths as functions of α.

(d) Based on the diagram, make a conjecture about the locus of the highest points of the water jets. Prove your conjecture analytically.

(e) How does the parameter v_0 influence the curve computed in (d)? Discuss the result analytically.

(f) Determine v_0 for a fountain of radius 2 m in the apartment of the Newrich family.

Chapter 2
First Order Differential Equations

2.1 Concepts and Definitions

Differential equations play a vital role in all fields of the natural and technical sciences, and increasingly also in economics and ecology. They describe deterministic processes or models thereof. They are equations in some **unknown function** of one or several variables such that the unknown function as well as its derivatives may occur.

In this book, we shall restrict ourselves to **ordinary differential equations**, that is, the unknown function depends on one single variable only. This is a rather drastic restriction. For example, wave processes, diffusion processes, and the Maxwell equations (which describe the complete theory of electrophysics) are represented by partial differential equations[1]. Nevertheless, the topic of ordinary differential equations is quite demanding. In practical cases, it is often impossible to find analytical solutions. However, numerical approximation methods are at hand.

Example 2.1 Let us show that the differential equation

$$y(t)^2 \cdot y'(t) = t^2$$

has the solution $y(t) = \sqrt[3]{t^3 + C}$ where C is an arbitrary constant:
$$y'(t) = [(t^3 + C)^{1/3}]' = \frac{1}{3}(t^3 + C)^{-2/3} \cdot 3t^2 \quad \text{and} \quad y(t)^2 = (t^3 + C)^{2/3}.$$

The product of the two expressions indeed equals t^2. ◇

---

[1] Partial differential equations are differential equations for functions of several variables. For instance, the one-dimensional wave equation for the unknown function $u(x,t)$ with x = position and t = time is $\frac{\partial^2 u}{\partial t^2} = c^2 \frac{\partial^2 u}{\partial x^2}$.

A. Fässler, *Fast Track to Differential Equations*,
https://doi.org/10.1007/978-3-030-83450-0_2

2.2 Geometrical Aspects, Direction Fields, Isoclines

The general differential equation of the first order has the following form:

$$y' = f(t, y).$$

Here the given function $f(t, y)$ varies with two independent variables t and y.
A solution $y = y(t)$ has the property that it satisfies the equation $y'(t) = f(t, y(t))$
identically **for all t-values on some given interval**.

We may visualize the differential equation by a **direction field**, because $f(t, y)$ provides the slope of any solution curve passing through the point (t, y). Hence the
direction field can give us some valuable qualitative insight into the solutions.

At each point the solution has the slope given by the direction field. The graphs of
different solutions cannot intersect each other, because at each point the slope is
unique.

In geometrical problems it is often customary to use the independent variable x
rather than the variable t:

$$y' = f(x, y).$$

When discussing the geometric behavior, it is often convenient to first determine the
so-called **isoclines**. Those are curves along which the slopes are constant. Hence
they satisfy the equation

$$f(x, y) = m$$

for different m-values, where m is the slope on the associated isocline.

Example 2.2 Let us analyze the solutions of the differential equation $y' = -\dfrac{x}{y}$.
The isoclines are described by

$$f(x, y) = -\frac{x}{y} = m \quad \Rightarrow \quad y - -\frac{1}{m} \cdot x.$$

Hence the isoclines are straight lines through the
origin with slopes $m^* = -\frac{1}{m}$, and so they are perpendicular to the tangents of the solution curves.
Therefore, the set of solution curves are concentric
circles about the origin.

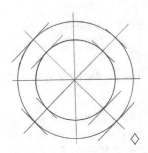

$\Diamond$

Example 2.3

Let us now discuss the solutions of the differential equation

$$y' = 1 + x - y.$$

Here the isoclines associated with the slope m are described by

$$1 + x - y = m \quad \Rightarrow \quad y = x + (1 - m).$$

These are parallel lines with slope 1 and
y-intercepts $1 - m$.

m	$1 - m$
0	1
1	0
2	−1
3	−2
−1	2
−2	3

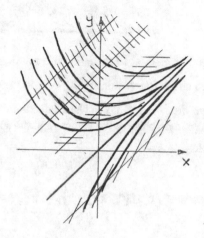

The figure shows the behavior of the
solutions. Obviously, a special solution
is $y = x$.   ◇

Example 2.4 In the special case of an **autonomous differential equation**

$$y' = f(y),$$

the right-hand side depends on y alone. Here the isoclines $f(y) = m$ are horizontal
lines. The direction field and thus also the graphs of the solutions are invariant under
horizontal translations.   ◇

Example 2.5 Logistic Differential Equation[2]

The following is an example of a logistic differential equation:

$$y' = 4y(1 - y).$$

For a constant y-value, the slope is $m = 4y(1 - y)$.

For $y = 1$ and $y = 0$, the slopes are $= 0$. Hence they are also solutions.

For $y = \frac{1}{2}$, the slope is 1.

Moreover, the slopes for $y = a$ and $y = 1 - a$ are the same, because $f(y)$ is a quadratic
function with axis of symmetry $y = \frac{1}{2}$.

---

[2] A general discussion follows in Exercise 66 of Chapter 3.

The following figure displays the typical S-shape of the solution curves. Also, they are invariant under translations and have a point symmetry with respect to $y = 1/2$ in the range $0 < y(t) < 1$.

$\diamondsuit$

Another method to visualize the solutions of the differential equation $y' = f(x,y)$ is to attach the vectors $\begin{pmatrix} 1 \\ f(x,y) \end{pmatrix}$ to various points (x,y), because the points (x,y) have the slope $f(x,y)$. The Mathematica command StreamPlot$[\ldots]$ generates a diagram displaying the vectors and the solution curves.

Example 2.6

$$y' = -\frac{1}{x} \cdot y + x. \tag{2.1}$$

StreamPlot$[\{1,\texttt{x-y/x}\},\{\texttt{x},\texttt{-2-2}\},\{\texttt{y},\texttt{-2,4}\}]$ furnishes this diagram:

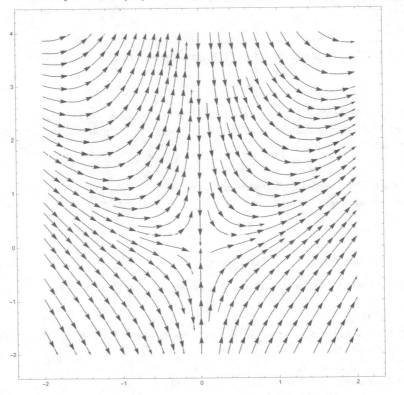

$\diamondsuit$

2.3 Linear First Order Differential Equations

Here $f(t,y)$ is a linear function of the unknown variable y. Hence the differential equation may be written in the following standardized form:

$$y' + a(t) \cdot y = b(t) \tag{2.2}$$

where $a(t)$ and $b(t)$ may be any complicated given functions.

We can devote ourselves to this problem only after we have solved the homogeneous case $b(t) = 0$.

2.3.1 Homogeneous Case

$$y' + a(t) \cdot y = 0. \tag{2.3}$$

The set of solutions is obtained as follows:

$$\frac{y'}{y} = -a(t) \implies \int \frac{y'}{y} \cdot dt = -\int a(t) dt \implies \ln|y(t)| = -A(t) + C,$$

where A is any antiderivative of a.

We exponentiate both sides and obtain

$$|y(t)| = e^{C - A(t)} = e^C \cdot e^{-A(t)} = k \cdot e^{-A(t)} \text{ where } k > 0.$$

Hence the set of solutions is

$$y(t) - k \cdot e^{-A(t)} \text{ with } k \in \mathbf{R}.$$

Here $y = 0$ is also a solution.

The nonzero solutions only differ by a multiplicative constant. For the **entire set of solutions** we also use the notion of the **general solution**.

With regard to Equation (2.2) the homogenized equation (2.3) is also called the **complementary equation**.

We verify the solutions by way of the Chain Rule:

$$y' = [k \cdot e^{-A(t)}]' = k \cdot e^{-A(t)} \cdot (-a(t)) = -a(t) \cdot y(t).$$

For the so-called **initial value problem (IVP)**

$$\left. \begin{array}{l} y' + a(t) \cdot y = 0 \\ y(t_0) = y_0 \end{array} \right\}$$

the unique solution for an arbitrary initial value y_0 is

$$y(t) = y_0 \cdot e^{-\int_{t_0}^{t} a(\tau) d\tau} = y_0 \cdot e^{-[(A(t) - A(t_0))]}.$$

Example 2.7 The IVP $\quad y' - 2ty = 0, \qquad y(0) = y_0$
has the solution $y(t) = y_0 \cdot e^{t^2}$, because $A(t) = -t^2$.   ◇

Example 2.8 Let us solve the homogeneous linear differential equation

$$y' + \frac{2t}{1+t^2}y = 0.$$

$$a(t) = \frac{2t}{1+t^2} \Rightarrow A(t) = \ln(1+t^2) \Rightarrow e^{-\ln(1+t^2)} = [e^{\ln(1+t^2)}]^{-1} = (1+t^2)^{-1}.$$

Hence the general solution is

$$y(t) = \frac{C}{1+t^2}. \tag{2.4}$$

The following diagram shows some of the solutions, which constitute an "onion".

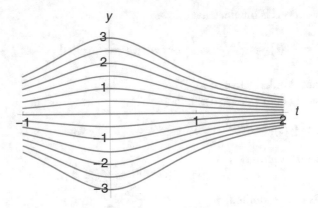

A basic statement: It is an illusion to believe that solutions of differential equations are always represented by a closed elementary expression. Most of the time, this is not the case. The reason for this is the integration, see Section 1.2.3.
In these cases one has to resort to numerical methods, which will be discussed later. Indeed, one has to pay a high price: parameters must not occur! But rather one has to look at a great number of graphs, in which the parameters are chosen numerically.

Example 2.9 The IVP

$$y' + \sqrt{1+t^2} \cdot e^{-t}y = 0, \qquad y(2) = 4$$

has the solution

$$y(t) = 4 \cdot \exp\left(-\int_2^t \sqrt{1+s^2} \cdot e^{-s} \cdot ds\right).$$

The integral cannot be expressed in a closed elementary manner.   ◇

2.3.2 Nonhomogeneous Case

The **method of** the so-called **integrating factor** allows us to solve the differential equation (2.2):
We multiply by some function $\mu(t)$, which has to be determined later:

$$\mu(t) \cdot y' + \mu(t) \cdot a(t) \cdot y = \mu(t) \cdot b(t). \tag{2.5}$$

If we could choose $\mu(t)$ such that $[\mu \cdot y]' = $ the left-hand side of (2.5), then we should be in the position to integrate the differential equation.
According to the Product Rule

$$[\mu \cdot y]' = \mu y' + \mu' y.$$

Hence

$$\mu(t)y' + \mu(t)'y = \mu(t)y' + \mu(t)a(t)y.$$

Subtracting $\mu(t)y'$ yields

$$\mu' = a(t)\mu.$$

This is a homogeneous linear differential equation in $\mu(t)$. It suffices to choose any arbitrary solution, for instance,

$$\mu(t) = e^{A(t)}, \tag{2.6}$$

where $A(t)$ is any antiderivative of $a(t)$.

By use of (2.6), the differential equation (2.5), being equivalent to (2.2), becomes

$$[\mu(t) \cdot y]' = \mu(t) \cdot b(t).$$

Integrating both sides yields

$$\mu \cdot y = \int \mu(t) \cdot b(t) \cdot dt + C.$$

Dividing both sides by $\mu(t) = e^{A(t)}$ results in the set of all solutions, also called the **general solution of the first order nonhomogeneous differential equation**:

$$y(t) = e^{-A(t)} \cdot \int e^{A(t)} \cdot b(t) \cdot dt + C \cdot e^{-A(t)}.$$

If in particular we choose the constant $C = 0$, then we find that the first term is a special solution of the nonhomogeneous differential equation, also called a **particular solution** $y_p(t)$.
Obviously, the following holds:

Theorem 10.

(a) *The general solution of the nonhomogeneous linear differential equation*

$$y' + a(t)y = b(t)$$

is the sum of a particular solution and the general solution of the complementary equation (i.e. the set of all solutions of the homogeneous case $y' + a(t)y = 0$):

$$y(t) = y_p(t) + C \cdot e^{-A(t)}.$$

Here $y_p(t)$ is any solution and $A(t)$ any antiderivative of $a(t)$.

(b) *Imposing an initial condition on the linear differential equation makes the solution y unique, provided the given functions a and b are at least piecewise continuous. In this case, y is continuous and has corners wherever a and b have isolated jumps.*

Remark:
Piecewise continuous functions arise in switch-on and switch-off processes.

Example 2.10 Let us solve the differential equation $y' + \dfrac{2t}{1+t^2} \cdot y = \dfrac{1}{1+t^2}$.

$$e^{A(t)} = e^{\ln(1+t^2)} = 1 + t^2.$$

$$y(t) = \frac{1}{1+t^2} \cdot \int (1+t^2) \cdot \frac{1}{1+t^2} \cdot dt + \frac{C}{1+t^2} = \frac{t}{1+t^2} + \frac{C}{1+t^2}.$$

The particular solution $y_p(t) = \dfrac{t}{1+t^2}$ passes through the origin.

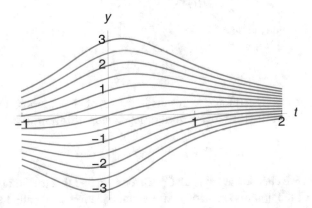

This figure illustrates the superposition of the particular solution y_p and the "onion" of general solutions of the complementary equation found in (2.4).   ◇

2.4 Existence and Uniqueness of Solutions

Without proof we state the following theorem, which is generally valid, also for nonlinear first order differential equations:

Theorem 11. *If in a (small) neighborhood of the point* (t_0, y_0) *the function* $f(t, y)$ *of the differential equation* $y' = f(t, y)$ *is*

- *continuous in both variables*
- *and its partial derivative* $\frac{\partial f(t,y)}{\partial y}$ *exists and is continuous,*

then in the neighborhood of (t_0, y_0) *there is a unique solution passing through the point.*

Remark: The partial derivative $\frac{\partial f(t,y)}{\partial y}$ is the derivative with respect to the variable y, while t is held constant. For instance, for $f(t, y) = \sin t \cdot y^2$ we have $\frac{\partial f(t,y)}{\partial y} = 2y \sin t$.

Example 2.11 Let us consider the differential equation

$$y' = \sqrt{y}.$$

At every point $(C, 0)$ the two solutions $y_1(t) = 0$ and $y_2(t) = \frac{1}{4}(t - C)^2$ branch out. This does not violate the Existence and Uniqueness Theorem, because the function

$$\frac{d}{dy}\sqrt{y} = \frac{1}{2\sqrt{y}}$$

is discontinuous at all points where $y = 0$. It is said to have a singularity there. But through all points where $y > 0$ there is a locally unique solution of the differential equation (without initial condition), namely the corresponding small segment of the parabola. ◇

Example 2.12 The differential equation

$$y' = y^2$$

has the solutions

$$y(t) = \frac{1}{C - t} \text{ and } y = 0.$$

This confirms the Existence and Uniqueness Theorem: Through every point (t_0, y_0) with $y_0 \neq 0$ there is a unique solution such that $C = \frac{1 + t_0 \cdot y_0}{y_0}$. If $y_0 = 0$, then the solution is $y = 0$. ◇

Example 2.13 The differential equation

$$y' + \tan t \cdot y = \sin t \cos t$$

has the solutions

$$y(t) = C \cdot \cos t - \cos^2 t,$$

which may be generated by the Mathematica command

```
DSolve[y'[t]+Tan[t] y[t]==Cos[t] Sin[t],y[t],t]
```

Here is a plot of the three solutions with $C = -1, 0, 1$ (from bottom to top):

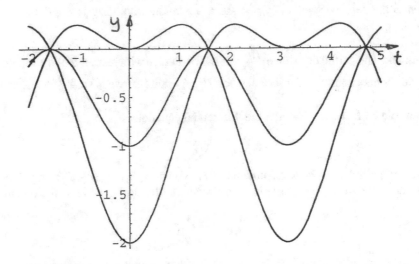

As

$$\frac{\partial f}{\partial y} = -\tan t \text{ for } f(t,y) = -\tan t \cdot y + \sin t \cos t,$$

there are singularities (i.e. points of discontinuities) at exactly $t = \pm \pi/2, \pm 3\pi/2, \ldots$.
At those points infinitely many solutions branch out.   ◇

Example 2.14
The differential equation

$$y' + \frac{1}{t}y = \frac{1}{t^2}$$

has the solutions

$$y(t) = \frac{\ln t}{t} + \frac{C}{t}.$$

Differential equation and solutions all are singular at $t = 0$.   ◇

Example 2.15

The differential equation

$$y' + \frac{1}{\sqrt{t}}y = e^{\sqrt{t}/2}$$

has the solutions

$$y(t) = \frac{4}{25}(5\sqrt{t} - 2)e^{\sqrt{t}/2} + \frac{C}{e^{2\sqrt{t}}}.$$

The differential equation is singular at $t = 0$, but the solutions are nonsingular. Here is a plot for $C = -1, 0, 1$:

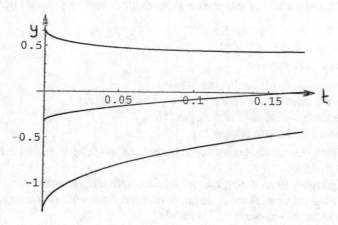

This is consistent with the Existence and Uniqueness Theorem. The reason is that the assumptions are sufficient, but not necessary, as is illustrated by this example. ◊

2.5 Separable Differential Equations

The differential equation

$$y' = f(t,y)$$

is **separable**, if f can be expressed as some product or quotient of a function in t and a function in y, i.e., if it is of the form

$$y' = \frac{g(t)}{h(y)}.$$

Example 2.16

$y' = \dfrac{a(b+y^3)}{\sin(2t)}$ is separable, $\qquad$ $y' = t - y^2$ is not separable. $\qquad$ ◊

Separable differential equations are integrable:

$$h(y) \cdot y' = g(t) \implies \int h(y) \cdot y' \cdot dt = \int g(t) \cdot dt + C.$$

Let H be any antiderivative of h and G any antiderivative of g. Then

$$H(y(t)) = G(t) + C$$

because of the Chain Rule $\dfrac{dH(y(t))}{dt} = h(y(t)) \cdot y'(t).$

In retrospect, our method of integration justifies the following **formal approach:**

$$\frac{dy}{dt} = \frac{g(t)}{h(y)} \quad \Longleftrightarrow \quad \int h(y) dy = \int g(t) dt + C.$$

The procedure is as follows:

(a) Use the notation dy/dt for the derivative,
(b) separate variables (no t's on the left, no y's on the right),
(c) integrate the left-hand side with respect to y,
     the right-hand side with respect to t.

The above indicates that this approach is correct, although we integrate with respect to different variables.

A word of caution: Even if integrating the functions h and g is not a problem, which often isn't even the case, there is no guarantee that the resulting ordinary equation can be solved for $y(t)$ by analytical methods.

As we shall learn later, a convenient alternative is to use numerical methods to solve such differential equations.

Example 2.17 Let us consider the nonlinear differential equation

$$e^y \cdot y' - t - t^3 = 0 \quad \implies \quad e^y \cdot \frac{dy}{dt} - t - t^3 = 0.$$

Separating variables and integrating both sides yields

$$\int e^y dy = \int (t + t^3) dt + C \quad \implies \quad e^y = \frac{1}{2}t^2 + \frac{1}{4}t^4 + C$$

leading to the solutions $y(t) = \ln(\dfrac{1}{2}t^2 + \dfrac{1}{4}t^4 + C).$

$\diamond$

Example 2.18 Let us consider the nonlinear IVP
$$y' = 1 + y^2 \qquad \text{with} \quad y(0) = 1.$$

Separating variables and integrating both sides yields

$$\int \frac{1}{1+y^2} dy = \int dt + C \quad \implies \quad \arctan y = t + C \quad \implies \quad y(t) = \tan(t + C).$$

The initial condition

$$y(0) = \tan(C) = 1 \Rightarrow C = \frac{\pi}{4}$$

leads to the solution $y(t) = \tan\left(t + \frac{\pi}{4}\right)$ where $t \in \left(-\frac{3\pi}{4}, \frac{\pi}{4}\right)$,

which is represented by a branch of the tangent function with a zero at $t = -\frac{\pi}{4}$.
Thus the solution exists only on a bounded open interval.   ◇

Example 2.19   Let us consider the separable nonlinear IVP

$$y \cdot y' = (1 + y^2)\sin t \qquad \text{with} \quad y(0) = -1.$$

$$\int \frac{y}{1 + y^2} \cdot dy = \int \sin t \cdot dt + C \implies \ln(1 + y^2) = -2\cos t + C.$$

Exponentiating both sides gives

$$1 + y^2 = e^{-2\cos t + C} = e^C \cdot e^{-2\cos t} = k \cdot e^{-2\cos t} \text{ where } k > 0.$$

Hence $y(t) = \pm\sqrt{k \cdot e^{-2\cos t} - 1}$.
The initial condition suggests a negative solution:

$$y(0) = -1 = -\sqrt{ke^{-2} - 1} \Rightarrow ke^{-2} = 2 \Rightarrow k = 2e^2.$$

Therefore,

$$y(t) = -\sqrt{2 \cdot e^{2(1 - \cos t)} - 1}.$$

As $1 \le e^{2(1 - \cos t)} \le e^4$, the solution exists on all of the real number line.

It is 2π-periodic and oscillates between -1 and $-\sqrt{2 \cdot e^4 - 1} \approx -10.4$.   ◇

Example 2.20   Now we show that the nonlinear differential equation

$$y' = \frac{y(-v + w \cdot x)}{x(+a - b \cdot y)} \qquad \text{where} \quad x, y > 0$$

is separable, but not explicitly solvable for y.
The constant parameters a, b, v, w are all > 0. Here x is the independent variable. In the numerator we factor out x, in the denominator we factor out y, then we remove the common factor xy to obtain

$$y' = \frac{-\frac{v}{x} + w}{+\frac{a}{y} - b} \implies \int \left(\frac{a}{y} - b\right) \cdot dy = \int \left(-\frac{v}{x} + w\right) \cdot dx + C.$$

$$a \cdot \ln y - b \cdot y = -v \cdot \ln x + w \cdot x + C \implies C = a \cdot \ln y - b \cdot y + v \cdot \ln x - w \cdot x.$$

This cannot be solved explicitly.
If we regard the right-hand side as a function $f(x, y)$ in the two variables x and y, then its graph is a spatial surface over the (x, y)-plane with $z = f(x, y)$. The solutions of the differential equation for different C-values correspond to the contours of the surface at altitudes $z = C$.
In Section 4.11 (predator-prey problem) the solutions are displayed graphically. ◇

Example 2.21 Integral Equation.

Find the functions $y(x)$ such that for every x the area under the curve is c times that of the rectangle in the diagram, i.e. the requirement is

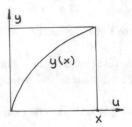

$$\int_0^x y(u) \cdot du = c \cdot x \cdot y(x).$$

By differentiating this integral equation, while applying the Product Rule to the right-hand side, we arrive at

$$y(x) = c[y(x) + x \cdot y'(x)].$$

Solving for y' and separating variables results in

$$y' = \frac{1-c}{cx}y \quad \Longrightarrow \quad \int \frac{dy}{y} = \int \frac{1-c}{cx}dx + k.$$

Integration and subsequent exponentiation produces the solution

$$\ln y = \frac{1-c}{c}\ln x + k = \ln x^{(1-c)/c} + k \quad \Longrightarrow \quad y(x) = \lambda x^{\frac{1-c}{c}}.$$

We recommend to check this solution by substituting it into the integral equation.

Here are some special cases:
(a) $c = 1$: constant functions $y(x) = \lambda$,
(b) $c = 1/2$: linear functions $y(x) = \lambda x$,
(c) $c = 2/3$: radical functions $y(x) = \lambda \sqrt{x}$,
(d) $c = 2$: singularity at $x = 0$. $y(x) = \lambda \frac{1}{\sqrt{x}}$. Checking the solution here is particularly interesting, because it involves an improper integral.

2.6 Autonomous Differential Equation and Stability

An autonomous differential equation $y' = f(y)$ has the property that the right-hand side is not an explicit function of the variable t.

The direction field of an autonomous differential equation is invariant under translations in horizontal direction. The isoclines are horizontal lines.

Every zero of the function $f(y)$ is a constant solution of the differential equation. Such a solution is called an **equilibrium point or a critical point.**

In the following example we inquire into the stability of constant solutions (equilibrium points).

Example 2.22 Let us discuss the solutions of the differential equation

$$y' = y(y-1)(y-2)$$

with the aid of the direction field:

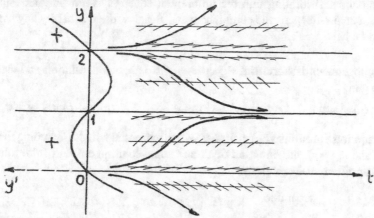

The curve on the left is the graph of $y' = y(y-1)(y-2)$ in (y, y')-coordinates. What happens if we start near an equilibrium point, say, due to a small perturbation?

It turns out that this example has two very different types of equilibrium points:

- A very small perturbation results in a solution that "escapes" the equilibrium point: **unstable or repelling equilibrium point.**
- If the perturbation is not too large, then the solution converges to the equilibrium point: **asymptotically stable or attracting equilibrium point**.

The points $y = 0$ and $y = 2$ are unstable (repelling), point $y = 1$ however is asymptotically stable (attracting).

Remark: In Section 4.11 a more general concept of stability will be discussed.

$\Diamond$

We may generalize our findings:

- A zero of $f(y)$ with negative (positive) slope corresponds to a stable (unstable) equilibrium point.
- If the slope of a zero equals 0, then it corresponds to a **semi-stable equilibrium point**. That is, the solutions on one side of the equilibrium point are attracted by it, while the solutions on the other side of the equilibrium point are repelled by it.[3]

---

[3] The notion of semi-stability is also used for limit cycles of solutions. In the interior and exterior of a limit cycle the solutions behave entirely differently: they are either attracted or repelled by the limit cycle.

2.7 Exercises Chapter 2

As a rule, the following exercises should be solved by hand. Checking by differentiating and susbstituting into the differential equation is not only an important process, but also deepens the comprehension. Besides, there is hardly any software without faults!

33. Homogeneous Differential Equations. Give a general solution and check by plugging in:

$$\text{(a)} \ y' = a \qquad\qquad \text{(b)} \ y' + k \cdot y = 0 \qquad\qquad \text{(c)} \ y' + (\sin t) \cdot y = 0$$

34. Graphing Solutions via Isoclines. Sketch various isoclines (including those for $m = 0$ and $m = \infty$), the direction field, and some solution curves of the following nonlinear differential equations:

(a) $y' = \dfrac{x}{y}$ \qquad separable.

(b) $y' = x^2 + y^2$ \quad not separable.

(c) $y' = \dfrac{y-x}{y+x}$ \quad not separable.

35. Nonhomogeneous Differential Equations. Compute the general solution of the nonhomogeneous differential equation and check your answer by substituting:

$$\text{(a)} \ y' - 2xy = x \qquad \text{(b)} \ xy' + y + 4 = 0 \qquad \text{(c)} \ xy' + y - \sin x = 0$$

36. Discontinuous Initial Value Problem. Compute the continuous solution of the IVP

$$y' + a(t) \cdot y = 0 \quad \text{where } y(0) = 1$$

for the piecewise continuous function

$$a(t) = \begin{cases} -1 & \text{if } 0 \le t \le 1 \\ 0 & \text{if } t > 1 \end{cases}.$$

37. Uniqueness. Analyze existence and uniqueness of the solutions of $y' = y$.

38. Singularities. Verify by substitution that the differential equation

$$xy'(x) + y(x) = x^2$$

with singularity at $x = 0$ has the solutions

$$y(x) = \frac{x^2}{3} + \frac{C}{x}.$$

Show that through each given point $(a \neq 0,\ b)$ there is a unique solution. In particular, determine the integration constant C.
Interestingly, there also exists a unique solution through the origin, which one? Is this statement consistent with the Existence and Uniqueness Theorem?

39. Branching of Solutions.
(a) Verify by substitution that

$$y_1(t) = -\frac{1}{4}t^2 \qquad \text{and} \qquad y_2(t) = 1-t$$

are solutions of the nonlinear differential equation

$$y' = \frac{1}{2}(-t + \sqrt{t^2 + 4y}).$$

(b) Show that the two solutions have a point of tangency B and compute its coordinates.

(c) Show that B, indeed, violates at least one of the assumptions of the Existence and Uniqueness Theorem.

40. Infinitely Many Solutions.
Compute the entire set of solutions of the differential equation $y' = \frac{y}{x}$. Explain why there are several (here even infinitely many) solutions passing through the origin. Doesn't this violate the Existence and Uniqueness Theorem?

41. Verifying Solutions.
Verify that $\quad y(t) = t \cdot (\ln t + C)^2 \quad$ is the general solution of

$$y' = \frac{y}{t} + 2\sqrt{\frac{y}{t}}.$$

Hint: When differentiating, make use of the Product and the Chain Rules.
Remark: All solutions pass through the origin, because $\lim\limits_{t \to \infty} t \cdot (\ln t)^2 = 0$.

42. Nonlinear Initial Value Problem.
Solve by use of a calculator or a computer

$$y' = y \cdot (1-y) \cdot x \quad \text{where } y(0) = 2.$$

43. Stability.
Let the differential equation

$$y' = y(1-y)$$

be given.

(a) Sketch the direction field and the solution curves.

(b) Analyze the stability of the critical points.

(c) Find the solution satisfying the initial condition $y(0) = 2$.

44. Initial Value Problem.
Compute the solutions of the IVP

$$\left. \begin{aligned} y' &= x^3 y^2 \\ y(2) &= 3 \end{aligned} \right\}.$$

45. Neither Linear Nor Separable.
Convince yourself that the differential equation $y' = t - y^2$ is neither linear nor separable.

46. No Closed Elementary Form. Convince yourself that the differential equation

$$y' = \frac{1}{y + \sin y}$$

can be integrated, but the resulting equation cannot be solved in a closed elementary manner.

47. Geometrical Problem. Determine a curve $y(x)$ such that the tangent line at $x = a$ intersects the x-axis at $x = a/2$. This property is to hold for all a. Check your answer.

48. Equilibrium Points. Investigate the equilibrium points of the differential equations

(a) $y' = f(y) = my + b$ and   (b) $y' = (y - 2)(y - 3)^2$.

Chapter 3
First Order Applications

Let us consider a quantity $f(t)$ that varies with time. Its change per unit time, i.e. the quotient $\frac{\Delta f}{\Delta t}$, is called the **average rate of change** during the time period Δt.

The difference quotient $\frac{\Delta f}{\Delta t}$, as $\Delta t \to 0$, converges to the differential quotient $\frac{df}{dt}$ of the function f (also called the derivative), which describes the **instantaneous rate of change** of f. Depending on the dimension of the quantity f, we call it the mass flow rate, the rate of change in displacement = instantaneous velocity, the temperature change rate, the growth rate, etc.

Often a quantity $f(t)$ changes at a rate **proportional to the quantity $f(t)$ itself**. For small Δt, this is approximated by

$$\text{instantaneous rate of change} \approx \frac{\Delta f}{\Delta t} \approx k \cdot f(t).$$

Depending on whether the proportionality factor k is positive or negative, the quantity $f(t)$ increases or decreases.

As "instantaneous" means letting $\Delta t \to 0$, the above procedure is accurately described by

$$f' = k \cdot f.$$

The general solution was found to be the exponential function (see Theorem 4 of Section 1.1.3)

$$f(t) = Ce^{kt}.$$

The following four Sections 3.1–3.4 deal with the exponential behavior of real-life applications.

3.1 Population Model

If the growth rate of a population p is proportional to the population size, then

$$\frac{dp}{dt} = r \cdot p$$

where r is the constant relative growth rate.

If the population initially, at time $t = 0$, has size $p(0) = p_0$, then the solution is

$$p(t) = p_0 \cdot e^{rt}.$$

By population we mean, for instance, the number of human beings, fish, rabbits, bacteria. This model is attributed to Malthus.[1]

As long as the resources, such as food supply and habitat size, are ample, the model named after Malthus is justified. But as soon as the resources get scarce, the model fails. The reason is that a population cannot grow exponentially over an extended period of time.

The so-called logistic growth model, investigated in Exercise 66, will take into account the limitations of resources.

3.2 Newton's Law of Cooling

An object is placed into an environment (liquid bath or gas) to cool down. The difference $D(t) = T(t) - U(t)$ between the temperature $T(t)$ of the object and the temperature $U(t)$ of its surroundings varies with time according to the following law: The temperature difference $D(t)$ decreases at a rate that is proportional to its current value, i.e.,

$$D'(t) = -k \cdot D(t) \text{ where } k > 0.$$

Of course, the constant k depends on many factors (for instance, k in water is much larger than in air).

Cooling and warming processes in nature and technology are subject to this law.

We limit ourselves to the special case of $U = $ constant. This usually holds in practical situations where the reservoir of the surroundings is large. Then

$$D(t) = T(t) - U \Longrightarrow D'(t) = T'(t),$$

implying an autonomous linear differential equation for the function $T(t)$:

$$T' = -k(T - U).$$

---

[1] Thomas Robert Malthus (1766-1834) was an English economist.

A particular solution is the zero of the right-hand side: $T_p = U$.
Thus the general solution is

$$T(t) = U + C \cdot e^{-kt}.$$

The constant C is determined by the initial temperature $T(0) = T_0$. Thus the solution to the IVP is

$$T(t) = U + (T_0 - U)e^{-kt}.$$

The temperature approaches U exponentially.
Left to the reader as an exercise: This solution is also valid for warming processes, in which case $(T_0 - U) < 0$.

Example 3.23 Champagne should be consumed at a temperature of 7 °C. A bottle from the wine cellar of temperature 14 °C is set into a bucket of ice-and-water mixture. After half an hour it has cooled down to 10 °C. How long does it take until the champagne is ready to be served?

$$T(t) = 14 \cdot e^{-kt} \implies 10 = 14 \cdot e^{-k/2} \implies k/2 = \ln \frac{14}{10} \implies k = 0.67294.$$

The time t is obtained from the equation

$$7 = 14 \cdot e^{-kt} \implies e^{-kt} = 1/2 \implies kt = \ln 2.$$

With the k-value found above, we obtain $t = \ln 2/0.67294 = 1.030$ h.
Enjoy the champagne after about 1 hour! ◇

3.3 Radioactivity and Radiocarbon Dating

3.3.1 Radioactivity

Let $N(t)$ = number of isotopes of an element at time t. From statistical considerations we know that $N(t)$ decays at a rate proportional to the current number of isotopes. For small Δt, this leads to

$$\frac{\Delta N}{\Delta t} \approx \frac{dN}{dt} = -k \cdot N(t) \text{ where } k > 0. \qquad (3.1)$$

From the initial condition $N(0) = N_0$ we infer that

$$N(t) = N_0 \cdot e^{-kt}.$$

The factor k in the solution relates to the half-life of the material.

By **half-life** we mean the time $T_{1/2}$ until half of the radioactive material has decayed. The relationship between k and $T_{1/2}$ is obtained from taking logarithms on both sides:

$$e^{-kT_{1/2}} = 1/2 \implies -kT_{1/2} = \ln(1/2).$$

Or simply

$$k \cdot T_{1/2} = \ln 2. \tag{3.2}$$

3.3.2 Radiocarbon Dating

One of the most accurate methods for determining the age of archaeological finds is the so-called radiocarbon dating. It was developed around 1949 by Willard Libby[2]: The earth's atmosphere is constantly bombarded by cosmic radiation. The resulting neutrons react with nitrogen to form the carbon $C14$ isotope.

The latter is a radioactive substance with a half-life of 5,745 years and is absorbed by living organisms in the form of CO_2. At the time of death, the absorption of $C14$ ceases. As a consequence, the (relatively low and harmless) decay rate per gram and minute decreases.

Suppose that cosmic radiation over the millenia has been constant, then a living piece of wood has a constant decay rate per gram and minute at all times. By measuring the reduced decay rate of an excavated organic find, one can compute the age of the sample.

---

[2] Willard Libby (1908–1980), chemist and physicist from the U.S.A., was awarded the Nobel prize for chemistry in 1960 for his work on radiocarbon dating.

Example 3.24 Caves of Lascaux in France.
Charcoal found in the layers of the caves of Lascaux in 1950 showed a decay rate of
0.97 decays per minute and gram. Living wood, however, has a decay rate of 6.68
decays per minute and gram.
We wish to determine the age of the charcoal and hence the approximate age of the
famous paintings.
It is known that at time $t = 0$, when the living piece of wood was buried, the decay
rate per minute and gram was

$$\frac{dN(0)}{dt} = 6.68.$$

At time τ of the discovery, however, the charcoal had a decay rate per minute and
gram of

$$\frac{dN(\tau)}{dt} = 0.97.$$

According to (3.1), the ratio of the decay rates equals the ratio of the number of
current isotopes:

$$e^{-k\tau} = \frac{0.97}{6.68}. \tag{3.3}$$

Equation (3.2), together with the half-life $T_{1/2} = 5,745$ years of $C14$, implies

$$k = \frac{\ln 2}{T_{1/2}} = 0.00012065.$$

Taking logarithms on both sides of (3.3) yields

$$-k \cdot \tau = \ln\left(\frac{0.97}{6.68}\right) \Rightarrow \tau = 15,993.$$

Hence the paintings found in Lascaux date back to approximately 16,000 years ago.

3.3.3 More on Radiocarbon Dating

Since about 1978 the Accelerator Mass Spectrometry (AMS) is available, a more
precise radiocarbon dating method. It directly measures the ratio between the numbers of isotopes $C14$ and $C12$, not just the decay ratios. $C14$ is much rarer than $C12$,
namely the ratio in living materials is only about 10^{-12}. The measuring method of
the AMS is by far more accurate than the earlier method of counting radioactive
decays. On top of this, only small samples of the material are needed.

Example 3.25 The Turin Shroud is a cloth of linen 4.4 m long and 1.1 m wide. It bears the image of the front and back of a human being and is kept at the Turin Cathedral. The documented first mention of the cloth dates back to the 14th century. It remained in the possession of various noble families and the House of Savoy and was handed over to the catholic church only in the late 20th century.

Three independent AMS analyses were performed, where each time only about 12 mg of the cloth was required. The three laboratories came to the conclusion that the cloth dates from the following periods:

Arizona: 1273 - 1335 AD,   Oxford: 1170 - 1230 AD,
Zurich (ETH): 1250 - 1298 AD

Based on the three analyses the journal *Nature* [7] determined an average value of 0.91825, meaning that the proportion of C14 in dead material has dropped to 91.83% of the C14-proportion in living material.

Therefore, the age τ of the shroud is obtained as follows:

$$e^{-k\tau} = 0.91825 \quad \Longrightarrow \quad \tau = -\ln 0.91825/k = 707 \text{ years.}$$

The differing time periods are due to the inaccuracy of the measuring method. But also the half-life of C14 could be a reason for the deviations. It is now set to 5730 ± 40 years, while Libby worked with 5586 ± 30 years.

In the following, we show that the inaccuracy in the half-life compared to the inaccuracy in the measurements is practically negligible:

$$e^{-(\ln 2/5730)\cdot 707} = 0.91803, \quad e^{-(\ln 2/5770)\cdot 707} = 0.91857, \quad e^{-(\ln 2/5690)\cdot 707} = 0.91748.$$

However, a deviation of the measurement of the C14-proportion by about 1% results in a time difference of 100 years:

$$\tau_1 = \frac{-\ln 0.91}{\ln 2/5730} = 780, \qquad \tau_2 = \frac{-\ln 0.92}{\ln 2/5730} = 689. \qquad \diamond$$

Remark: The so-called bomb peak[3] in the period of 1960–1985 led to an increased ratio of about 20% to 70%, caused by nuclear weapons testing conducted aboveground. Therefore, when dating samples younger than 70 years, one has to account for these complicating effects[4].

The bomb peak violates the assumption that the ratio of C14/C12 was constant. For older samples this is not a problem, as they were not exposed to the atmosphere in the period of 1960–1985.

---

[3] There are also other effects, for example the so-called Suess effect, caused by industrialization.

[4] Example in [12].

3.4 Compound Interest

Example 3.26 Interest Compounded Continuously

If a principal earns an interest of $q\%$ compounded continuously, then the rate of change of the principal $K(t)$ is proportional to the current amount. Thus the differential equation is

$$\frac{\mathrm{d}K}{\mathrm{d}t} = \frac{q}{100}K.$$

With the initial principal $K(0) = K_0$ the solution becomes

$$K(t) = K_0 \cdot e^{\frac{q}{100}t}.$$

Question: What interest rate of $q\%$ compounded continuously is equivalent to an annual interest rate of $p\%$?

After one year,

$$K_0 \cdot e^{\frac{q}{100}} = K_0\left(1 + \frac{p}{100}\right).$$

Canceling K_0 and taking logarithms on both sides yields

$$\frac{q}{100} = \ln\left(1 + \frac{p}{100}\right).$$

Now we apply the approximation formula $\ln(1+x) \approx x - \frac{x^2}{2} + \frac{x3}{3}$ with $x = \frac{p}{100}$ and multiply by 100 to obtain a useful relationship between p and q:

$$q = p\left(1 - \frac{p}{200} + \frac{p^2}{30{,}000}\right).$$

This result is also valid for negative interest rates!

The following graph compares p and q:

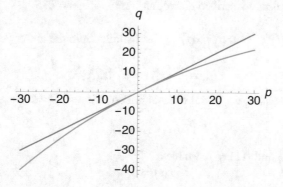

Two numerical examples: $(p/q) = (8/7.697), (-8/-8.049)$. ◇

3.5 Specified Elasticity Function in Economics

Earlier, we were given a function $f(x)$, and we computed the elasticity function $\varepsilon_{f,x}$ according to (1.9).

But we may also ask the converse question (see also [35]), which in practice is perhaps less significant: Find a function $f(x)$ for a given elasticity function $\varepsilon_{f,x}$. This leads to a homogeneous linear differential equation.

Example 3.27 Let

$$\varepsilon_{f,x} = a + \sqrt{bx}$$

with parameters a, b. Then

$$a + \sqrt{bx} = \frac{f'}{f} \cdot x \quad \Longleftrightarrow \quad f' = f \cdot \left(\frac{a}{x} + \sqrt{\frac{b}{x}} \right).$$

An antiderivative of the term between the parentheses is $\quad a \ln x + 2\sqrt{bx}$. Thus

$$f(x) = k \cdot x^a \cdot e^{2\sqrt{bx}}. \qquad\qquad \diamond$$

3.6 Evaporation of a Rain Drop

As is known, a rain drop is nearly of spherical shape. We assume that during the fall through the air, the rate of evaporation is proportional to the drop's surface area. Determine the radius $r(t)$ as a function of time, if in the beginning it is 3 mm and an hour later it is 2 mm.

Surface area of a sphere: $S = 4\pi r^2$ $\qquad$ Volume of a sphere: $V = \frac{4\pi}{3} r^3$

The physical process described above implies $\qquad \dfrac{dV}{dt} = k \cdot S.$

$$\frac{dV}{dt} = \frac{d}{dt}\left(\frac{4\pi}{3} r(t)^3 \right) = 4\pi r(t)^2 \cdot \dot{r}(t) \qquad \text{substituted into the above equation yields}$$

$$4\pi r(t)^2 \cdot \dot{r}(t) = k \cdot 4\pi r(t)^2.$$

Cancelation results in the trivial differential equation $\dot{r}(t) = k$. Its general solution is

$$r(t) = kt + b.$$

Due to $r(0) = 3$ and $r(1) = 2$, we have
$$r(t) = 3 - t.$$

Thus the life span is 3 hours.

3.7 Mixing Problem

A cylindrical tank of volume 240 l is initially filled with a solution that contains S_0 kg salt dissolved in water. From time $t = 0$ on, 2 l/min of brine with the concentration of 0.10 kg salt/l flows through an inflow into the tank (see figure).
It is also known that 2 l/min of the well-stirred brine flows out.
Find the amount of salt $S(t)$ in kg.

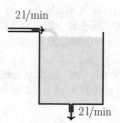

Flow budget:

- Entering amount: $2 \cdot 0.10 = \frac{1}{5}$ kg salt/min.
- Leaving amount: $\frac{2}{240} \cdot S(t)$ kg salt/min.

Net amount: $\quad \dfrac{\Delta S}{\Delta t} \approx \dfrac{1}{5} - \dfrac{1}{120} \cdot S(t).$

By letting $\Delta t \to 0$, we obtain the linear autonomous differential equation

$$\dot{S} = \frac{1}{5} - \frac{1}{120} \cdot S = f(S).$$

A particular solution is conveniently obtained as follows:

$$f(S) = 0 \implies S_\infty = 24 \text{ kg}.$$

With the set of solutions $S_h(t) = C \cdot e^{-\frac{1}{120}t}$ of the complementary equation, we obtain the general solution

$$S(t) = S_\infty + C \cdot e^{\frac{1}{120}t}.$$

The initial condition

$$S(0) = S_\infty + C = S_0 \implies C = S_0 - S_\infty.$$

Therefore, the solution is

$$S(t) = 24 + (S_0 - 24)e^{-\frac{1}{120}t}.$$

It increases or decreases exponentially towards 24 kg, depending on whether $S_0 < 24$ or $S_0 > 24$.

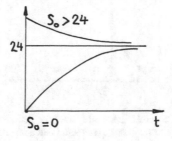

3.8 Vertical Launch of a Rocket Without Air Drag

A rocket of void mass m_0 kg and fuel mass m_1 kg before launch has a constant mass
flow rate (i.e. the burned fuel per unit time is constant) of μ kg/s. During the burning
phase it develops a constant thrust of
$$S = \mu \cdot r \text{ Newton}.$$
Here r is the constant relative exit velocity of the gas.
The equation of motion [5] for the climbing speed $v(t)$ during the burning phase is
$$(M - \mu \cdot t) \cdot \frac{dv}{dt} = S - (M - \mu \cdot t) \cdot g.$$
The term inside the parentheses on the left describes the mass as it decreases linearly
with time, and the term on the very right is the weight decreasing linearly with
time. Here the earth's gravitational acceleration g is assumed to be constant, and
$M = m_0 + m_1$ is the total mass before launch. Initial condition: $v(0) = 0$.
At the end of the burning phase, at time $t_1 = \dfrac{m_1}{\mu}$, maximum speed is attained, because until this moment the thrust S is in effect.

Moreover, during the burning phase, the resultant force on the right-hand side increases steadily, while the remaining mass on the left-hand side decreases steadily.
This results in a strictly monotonically increasing acceleration. The qualitative behavior of the speed $v(t)$ on the interval $0 \le t \le t_1$ is described by a curve that is
concave up, and so is that of the rising height $y(t)$.

After the burning phase ($t > t_1$), only the earth's constant gravity is at work:
The speed $v(t)$ decreases linearly with slope $-g$.

The qualitative behaviors of $v(t)$ and height $h(t)$ are represented in the following
diagram:

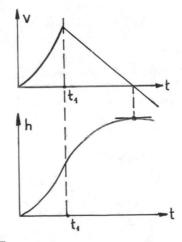

---

[5] Sir Issac Newton (1642–1726) created a world view for generations that was only limited in
its validity by Einstein's theory of relativity. He created calculus, classical mechanics, a theory of
light, the theory of gravity and a concept of determinism.

At maximum height h the velocity is $v = 0$.
Now we determine $v(t)$ for the burning phase by integrating

$$\frac{dv}{dt} = \frac{S}{M - \mu \cdot t} - g.$$

$$v(t) = \int \frac{S}{M - \mu \cdot t} dt - g \cdot t + C = -\frac{S}{\mu} \ln(M - \mu \cdot t) - g \cdot t + C.$$

$$v(0) = 0 = -\frac{S}{\mu} \ln M + C \implies v(t) = \frac{S}{\mu} [\ln M - \ln(M - \mu \cdot t)] - g \cdot t.$$

For the burning phase $t \le t_1$, we obtain

$$v(t) = \frac{S}{\mu} \cdot \ln \frac{M}{M - \mu \cdot t} - g \cdot t. \tag{3.4}$$

The model is flawed: For high speeds the air drag plays a critical role. At an altitude of 80 km the air density is only $1/10,000$th of that on sea level. The earth's gravitational acceleration g, however, decreases only by about 2%.

The present model is reasonable for small toy rockets (water/air with a small pump, about 40 cm high) and for the starting phase of large rockets.

The modeling problem 92 takes the air drag into account.

3.9 Gravitational Funnel

A gravitational funnel is a funnel-shaped surface made of metal with rotational symmetry. It is designed in a way such that a coin, kicked off tangentially at a reasonable initial speed and under the influence of the earth's gravity, spirals down the funnel while losing height only very slowly.

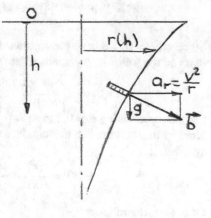

Let us compute the lateral curve $r(h)$ such that the coin at each height traces a circular path.
Because of the small friction, the path, as is desired, will take the shape of a tightly wound narrowing spiral.

First let us determine the total kinetic energy of a rolling coin (cylinder) with mass m, velocity v, radius ρ, and angular velocity $\omega = v/\rho$. The energy is a combination of translational and rotational energy. The moment of inertia is $J = \frac{1}{2} m \rho^2$.

$$E_{tot} = E_{trans} + E_{rot} = \frac{m}{2}v^2 + \frac{J}{2}\omega^2 = \frac{m}{2}v^2 + \frac{1}{4}m\rho^2\frac{v^2}{\rho^2} = \frac{3}{4}mv^2.$$

Conservation of energy means

$$\frac{3}{4}mv_0^2 + mgh = \frac{3}{4}mv^2 \quad\Longrightarrow\quad v^2 = v_0^2 + \frac{4}{3}gh,$$

implying that the velocity v increases with height h.

Now let us consider the resultant acceleration

$$\vec{b} = \vec{a}_r + \vec{g}.$$

Here $a_r = \dfrac{v^2}{r}$ is the magnitude of the radial acceleration and $\vec{g}$ is the acceleration of gravity.

When neglecting the friction, the tangent line of the curve $r(h)$ must be perpendicular to the resultant acceleration $\vec{b}$. Hence the slope $r'(h)$ satisfies

$$\frac{a_r}{g} = -\frac{1}{r'(h)} \quad\Longrightarrow\quad r'(h) = -\frac{g}{v^2/r} = -g \cdot \frac{r}{v^2}.$$

This leads to the linear homogeneous differential equation

$$r'(h) = \frac{-g}{v_0^2 + \frac{4}{3}g \cdot h} \cdot r(h).$$

An antiderivative of the fraction is $\quad A(h) = -\frac{3}{4}\ln(v_0^2 + \frac{4}{3}g \cdot h), \quad$ whence

$$r(h) = k \cdot e^{-\frac{3}{4}\ln(v_0^2 + \frac{4}{3}gh)} = k[e^{\ln(v_0^2 + \frac{4}{3}g \cdot h)}]^{-\frac{3}{4}} = k\frac{1}{(v_0^2 + \frac{4}{3}gh)^{3/4}}.$$

Since $r(0) = R$, we have the constant $k = Rv_0^{3/2}$. So we obtain, irrespective of the mass m, the following geometrical property:

$$r(h) = R\left(\frac{v_0}{v(h)}\right)^{3/2} \quad \text{where } v(h) = \sqrt{v_0^2 + \frac{4}{3}gh}.$$

For example, taking $R = 0.2$ m, $h = 0.4$ m, $v_0 = 1\,\frac{m}{s}$ results in $r(0.4) = 5.07$ cm. Now let us also consider the angular frequency with respect to the funnel's axis of symmetry:

$$\Omega(h) = \frac{v(h)}{r(h)} = \frac{v(h)^{5/2}}{R \cdot v_0^{3/2}} = \frac{1}{Rv_0^{3/2}}v(h)^{5/2}.$$

Of course, the relation

$$r(h) \cdot \Omega(h) = v(h)$$

holds. Hence the angular frequency increases with depth h faster than the radius decreases, because $v(h)$ increases with h.

3.10 Nuclear Waste Disposal by Use of Computer

Decades ago the American Nuclear Energy Agency disposed of concentrated radioactive waste by sealing it into containers and then dumping them into the sea at a site of depth 91.5 m.[6]

When ecologists questioned this procedure, the agency assured that the containers would not burst when hitting the seabed. Tests performed by engineers inferred that there would be a danger of the containers breaking up, if the impact speed exceeded 12 m/s.

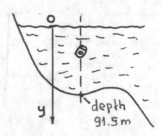

Data:

Total weight of a container: $G = 2347$ N,

Volume of a container: $V = 208$ l,

Density of sea water: $\rho = 1.025$ g/cm^3.

Several towing experiments carried out in water showed that the force of resistance W caused by the ocean current is approximately proportional to the velocity v:

$$W = c \cdot v \quad \text{where } c \approx 1.168 \ \frac{\text{N} \cdot \text{s}}{\text{m}}.$$

The position of the container relative to the towing direction affected c only slightly, so that we may work with the resistance force given above.

Newton's law of motion is responsible for describing the sinking process:

resultant force = mass · acceleration.

To reformulate this, we introduce the depth y according to the figure. If t is the time and A the buoyancy, then the velocity function $v(t) = \dot{y}(t)$ obeys

$$m \cdot \dot{v} = G - A - W.$$

As this problem is one-dimensional, we replace the vectors by real numbers. Vectors pointing down are positive, vectors pointing up are negative. With the buoyancy $A = -\rho V g$ we obtain

$$\dot{v} = \frac{1}{m}\left(G - \rho V g - cv\right) = f(v).$$

This is an autonomous linear nonhomogeneous differential equation.

From $f(v) = 0$, we conveniently obtain a particular solution

$$v_\infty = \frac{G - \rho V g}{c} = 219 \text{ m/s}.$$

---

[6] This problem is taken from [5], modified and converted into International Units (SI). With kind permission of the Springer Publishing Company.

Together with the complementary set of solutions, we obtain the general solution of the nonhomogeneous differential equation

$$v(t) = v_\infty + k \cdot e^{-\frac{c}{m}t}.$$

If in the most favorable case the initial velocity is $v(0) = 0$, then the velocity function reads

$$v(t) = v_\infty \cdot \left(1 - e^{-\frac{c}{m}t}\right) = 219 \cdot \left(1 - e^{-\frac{1}{204.8}t}\right)$$

with the following graph:

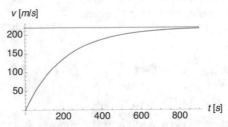

In regard of the limiting velocity v_∞, this model obviously does not make sense: For large velocities, the resistance force no longer depends linearly on the velocity.

In order to compute the impact velocity v_{floor} on the seafloor, we need the time t_{floor} of impact.

At this stage, let us first determine the depth function $y(t)$:

$$y(t) = \int_0^t v(s)\mathrm{d}s = v_\infty\left(s + \frac{m}{c}e^{-\frac{c}{m}s}\right)\Big|_0^t = v_\infty\left[t - \frac{m}{c}\left(1 - e^{-\frac{c}{m}t}\right)\right].$$

With the given constants, we obtain

$$y(t) = 219.2 \cdot \left[t - 204.8\left(1 - e^{-0.00488t}\right)\right].$$

The following graph shows the depth function $y(t)$, together with a marking of the depth of the seabed at 91.5 m. It allows us to read the time t_{floor} when the container touches the seafloor: $t_{\text{floor}} \approx 13.2$ s.

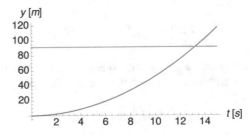

The associated velocity is

$$v(t_{\text{floor}}) \approx v(13.2) = 13.7 \text{ m/s}.$$

Hence there absolutely was a danger of the containers breaking apart!
Remark: In the meantime, the Nuclear Energy Agency has banned sea dumping of low-level radioactive waste.

3.11 Barometric Formulas

The aim is to determine the pressure $p(h)$ as a function of the altitude h. In the following models we shall consider the air as an ideal gas. If p is the pressure, V the volume, T the absolute temperature, then the mass m of the gas obeys the law

$$\frac{p \cdot V}{T} = \frac{m}{M} \cdot R = \text{constant},$$

where $M = 0.02896$ kg/mol is the average molar mass of atmospheric gases and $R = 8.314$ J/(K mol) the universal gas constant.

The law of ideal gas implies

$$\frac{m}{V} = \rho(h) = \frac{M}{R} \cdot \frac{p(h)}{T(h)}, \tag{3.5}$$

where $\rho(h)$ is the density of the gas as a function of the altitude h.

The following figure depicts a square column of air with height Δh and cross-sectional area A. The pressure at altitude h is p, at altitude $h + \Delta h$ it is $p + \Delta p$ (where $\Delta p < 0$).

Now we apply Newton's law to the air column at equilibrium state:

The force exerted by the pressure at the bottom is $p \cdot A$. It equals the force $(p + \Delta p) \cdot A$ at the top plus the weight G of the air column:
$$p \cdot A = (p + \Delta p) \cdot A + \rho \cdot \Delta h \cdot A \cdot g.$$

Here $\rho = \rho(h)$ is the air density as a function of the altitude h and g is the gravitational acceleration. Dividing by A yields
$$0 = \Delta p + \rho \cdot \Delta h \cdot g \quad \Longrightarrow \quad \frac{\Delta p}{\Delta h} = -\rho g.$$

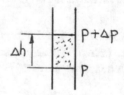

Taking the limit as $\Delta h \to 0$ yields the relationship between pressure and density:

$$\frac{dp}{dh} = -g \cdot \rho(h). \tag{3.6}$$

Substituting $\rho(h)$ by (3.5) leads to

$$\frac{dp}{dh} = -\frac{gM}{R} \cdot \frac{1}{T(h)} \cdot p(h).$$

With the pressure at sea level $p(0) = p_0$ as an initial condition, we obtain the general equation

$$p(h) = p_o \cdot \exp\left(-\frac{gM}{R} \int_0^h \frac{1}{T(s)} ds\right). \tag{3.7}$$

It is crucial to know the temperature curve $T(h)$ in order to compute the pressure $p(h)$. In the following, we shall consider different cases.

3.11.1 Isothermal Model

Here we assume that the temperature T is constant, admittedly a rough simplification (to be abandoned in the next model).

Owing to (3.5) and constant temperature, the density $\rho(h)$ is directly proportional to the pressure $p(h)$. Then (3.7) implies the **special barometric formula**, stating that the pressure decays exponentially with altitude h:

$$p(h) = p_0 \cdot e^{-\frac{gM}{RT} \cdot h} \quad \text{for} \quad T = \text{const.} \tag{3.8}$$

For $T = 15° C = 288 \ K$ and altitude h measured in meters, we have

$$p(h) = p_0 \cdot e^{-k \cdot h} \text{ where } k = 0.00011861.$$

For example, the pressure at altitude $h = \ln 2/k \approx 5840$ m is half the pressure p_0 at sea level.

3.11.2 Model for Linear Temperature Decrease

It is known that in the troposphere, i.e. up to an altitude of about 11 km, the temperature decreases nearly linearly with a gradient of $\gamma = 6.5$ K/km, i.e. $T(h) = T_0 - \gamma h$, with h measured in km. For the integral in the exponent of (3.7) we have

$$\int_0^h \frac{1}{T_0 - \gamma s} ds = -\frac{1}{\gamma}[\ln(T_0 - \gamma h) - \ln(T_0)] = -\frac{1}{\gamma} \ln\left(\frac{T_0 - \gamma h}{T_0}\right).$$

By substituting this into (3.7), we obtain the **barometric formula for linear temperature decrease (troposphere)**:

$$p(h) = p_o \cdot \left[\frac{T_0 - \gamma h}{T_0}\right]^{\frac{gM}{R\gamma}} \quad \text{where } \gamma = 6.5 \text{ K/km,}$$

T_0 is the absolute temperature and p_0 the pressure at sea level.

3.11.3 General Model

We want the integral in the exponent of (3.7) to make sense also for altitudes higher than 80 km. The standard atmosphere at an altitude between about 11 and 20 km is known to have a practically constant temperature. Up to about 50 km the temperature increases, and thereafter it decreases.

The Mean Value Theorem for integrals states that there is an average temperature T_m for an air column of height h satisfying

$$\frac{1}{T_m} \cdot h = \int_0^h \frac{1}{T(s)} ds.$$

Here T_m is the continuous harmonic mean of the temperature curve (see Exercise 52c), and as in (3.8) we obtain the **general barometric formula**

$$p(h) = p_0 \cdot e^{-\frac{gM}{RT_m}h}. \tag{3.9}$$

In the following, we show that the harmonic mean T_m hardly differs from the arithmetic mean: $\quad T_m \approx \frac{1}{h} \int_0^h T(s) ds.$

The reason is that the absolute temperature curve varies only slightly. As a real-life example, we may choose $T(h) = 293 - 6.5 \cdot h$ with $T(11 \text{ km}) = 221.5$ K:

$$\int_0^{11} \frac{1}{T(s)} ds = \int_0^{11} \frac{1}{293 - 6.5s} ds = -\frac{1}{6.5} \ln(293 - 6.5s)\Big|_0^{11} = \frac{1}{6.5} \ln(293/221.5) = 0.043038.$$
$$\frac{1}{T_m} \cdot 11 - 0.043038 \implies T_m = 255.6 \text{ K.}$$

In comparison, the arithmetic mean, due to linearity of T, is $\frac{293+221.5}{2} = 257.2$ K. Even for nonlinear temperature curves, the two means do not differ noticeably.

Remark: To compute the pressure curve $p(h)$ up to and above 11 km, it is reasonable to use the barometric formulas that are associated with the different altitudes.

3.12 Liquid Container

Let a cylindrical liquid container have height H and circular area A. The supply pipe and the drain hole have the same circular cross-sectional area a. The feed rate v from the supply pipe is constant.

Let us solve the following problems:

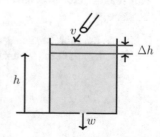

(a) Determine the differential equation for the time-varying water level $h(t)$.

(b) Compute $h_\infty = \lim_{t \to \infty} h(t)$.

(c) Find the inflow rates v for which the container doesn't overflow.

(d) Compute the function $h(t)$ for $v = 0$ and arbitrary initial water level $h(0) = h_0$.

(e) For the special case $v = 0$, compute and sketch the graph of the time $T(h_0)$ that passes until the container is empty. Here T is a function of the initial level h_0.

(f) Now consider the container with the following dimensions:

height $H = 6$ m, diameter $D = 5$ m, diameter of the drain hole $d = 5$ cm.

  (i) For $v = 0$, compute the emptying times $T(h_0)$ for the initial levels

  $h_0 : 1$ mm, 1 cm, 1 m, 6 m.

  (ii) Compute $h_\infty = \lim_{t \to \infty} h(t)$ for $v = 8$ m/s.

  (iii) Compute the maximum speed $v_{\max}$ for which the container does not overflow.

Solutions:

(a) Flow budget:

inflow rate: $v \cdot a$ m^3/s,

outflow rate: $w \cdot a$ m^3/s.

Hence for small Δt the time-dependent net volume $V(t) = A \cdot h(t)$ obeys

$$\frac{\Delta V}{\Delta t} = A \cdot \frac{\Delta h}{\Delta t} \approx a(v - w).$$

Conservation of energy for water particles of mass m (potential energy is converted to kinetic energy) implies

$$mgh = \frac{m}{2} w^2 \implies w(h) = \sqrt{2gh}.$$

Letting $\Delta t \to 0$ and dividing by A yields an autonomous nonlinear differential equation:

$$\frac{dh}{dt} = \frac{a}{A}(v - \sqrt{2gh}) = f(h).$$

(b)

$$f(h) = 0 \implies \sqrt{2gh} = v \implies h_\infty = \frac{v^2}{2g}.$$

This makes sense, because for h_∞ we have $v = w$: outflow and inflow rates are the same.

If $h(0) > h_\infty$, then the water level drops asymptotically towards h_∞,

if $h(0) < h_\infty$, then the water level rises asymptotically towards h_∞.

(c)

$$v \le \sqrt{2gH}.$$

(d)

$$\frac{dh}{dt} = -\frac{a}{A}\sqrt{2g} \cdot \sqrt{h} = -\lambda \cdot \sqrt{h},$$

$$\int \frac{dh}{\sqrt{h}} = -\int \lambda \cdot dt + C \Longrightarrow 2\sqrt{h} = -\lambda \cdot t + C \Longrightarrow \sqrt{h} = C - \frac{\lambda}{2} \cdot t.$$

Finally,

$$h(t) = \left(C - \frac{\lambda}{2} \cdot t\right)^2.$$

The initial condition

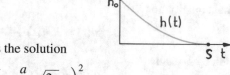

$$h(0) = h_0 = C^2 \Rightarrow C = \sqrt{h_0}$$

together with $\lambda = \frac{a}{A}\sqrt{2g}$ yields the solution

$$h(t) = \left(\sqrt{h_0} - \frac{a}{2A}\sqrt{2g} \cdot t\right)^2.$$

Its graph is a parabolic segment through point $(0, h_0)$ with vertex at $h = 0$.

(e) The emptying time is obtained from $h = 0$:

$$T(h_0) = \frac{2A}{a\sqrt{2g}} \cdot \sqrt{h_0}.$$

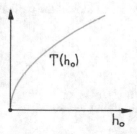

The graph of the radical function has a vertical tangent at $h_0 = 0$. This means that for small initial levels h_0, the emptying time increases extremely fast.

(f) (i) As $\frac{A}{a} = 10^4$, the emptying time for initial water levels h_0 is

$$T(h_0) \approx 4,515 \cdot \sqrt{h_0}.$$

For $h_0 = \{1 \text{ mm}, 1 \text{ cm}, 1 \text{ m}, 6 \text{ m}\}$, they are approximately $\{2.4 \text{ min}, 7.5 \text{ min}, 1.25 \text{ h}, 3.1 \text{ h}\}$.

(ii) Using the result in (b) gives $h_\infty = 3.26$ m.

(iii) Using the result in (b) gives $v_{\max} = 10.85$ m/s.

3.13 Electric Circuit

The figure depicts an electric circuit with alternating current voltage $u(t) = U\sin(\omega t)$, a resistor with resistance R, and an inductor with inductance L.

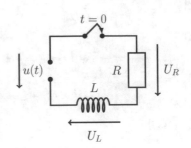

Find the current $i(t)$ after closing the switch at time $t = 0$.

We apply Kirchhoff's Law to the three voltages:

$$u(t) = U_R(t) + U_L(t).$$

This implies a linear nonhomogeneous differential equation for $i(t)$:

$$u(t) = R \cdot i(t) + L \cdot \frac{di}{dt}.$$

Division by L yields

$$\frac{di}{dt} + \frac{R}{L}i = \frac{U}{L} \cdot \sin(\omega t)$$

with the initial condition $i(0) = 0$.

The set of solutions of the complementary equation is given by $i_h(t) = k \cdot e^{-\frac{R}{L}t}$. Now let us look at the particular solution:

$$i_p(t) = \frac{U}{L}e^{-\frac{R}{L}t} \cdot \int e^{\frac{R}{L}t} \cdot \sin(\omega t) \cdot dt.$$

$$i_p(t) = \frac{U}{L}e^{-\frac{R}{L}t} \cdot \frac{RL}{R^2 + \omega^2 L^2}e^{+\frac{R}{L}t}\left[\sin(\omega t) - \frac{L\omega}{R}\cos(\omega t)\right].$$

Simplifying yields

$$i_p(t) = \frac{U}{R^2 + \omega^2 L^2}\left[R\sin(\omega t) - L\omega\cos(\omega t)\right].$$

The general solution of the nonhomogeneous problem is

$$i(t) = i_p(t) + k \cdot e^{-\frac{R}{L}t}.$$

The initial condition yields $0 = \dfrac{U}{R^2 + \omega^2 L^2} \cdot (-L\omega) + k \Longrightarrow k = \dfrac{UL\omega}{R^2 + \omega^2 L^2}.$

Therefore, the solution of the initial value problem is

$$i(t) = \frac{U}{R^2 + \omega^2 L^2}\left[R\sin(\omega t) - L\omega\cos(\omega t)\right] + \frac{UL\omega}{R^2 + \omega^2 L^2}e^{-\frac{R}{L}t}.$$

The first summand describes the **steady-state component** i_{st} of the solution. The second summand, subsiding exponentially, is called the **transient component** i_{tr} of the solution.

The total current rapidly turns into the steady state:

$$i(t) = i_{st}(t) + i_{tr}(t) \approx i_{st}(t).$$

A qualitative diagram illustrates this:

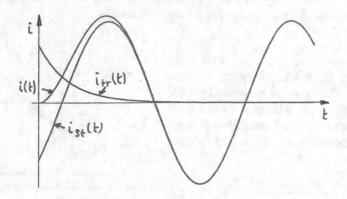

The trigonometric identity

$$a\cos(\omega t) + b\sin(\omega t) = A \cdot \sin(\omega t + \varphi) \qquad (3.10)$$

allows us to improve our understanding of the steady-state behavior.

For $i_{st}(t)$ we have $a = -L\omega \cdot k < 0, \quad b = R \cdot k > 0 \quad$ where $k = \dfrac{U}{R^2 + \omega^2 L^2}$,

hence the amplitude is $\qquad A = \sqrt{a^2 + b^2} = k\sqrt{R^2 + \omega^2 L^2} = \dfrac{U}{\sqrt{R^2 + \omega^2 L^2}}$

and the phase angle is $\qquad \varphi = \arctan(\ \frac{\omega L}{R}) \leq 0.$

The **phasor diagram** on the right visualizes the trigonometric identity above at time $t = 0$. The current $i_{st}(t)$ and the voltage $u(t)$ correspond to the projection of the phasors, i.e. of the vectors having lengths $\sqrt{a^2 + b^2}$ and U rotating with the angular frequency ω, onto the vertical axis.

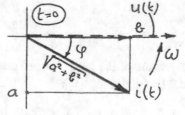

By taking $\alpha = \arctan\frac{\omega L}{R} \geq 0$ we may rewrite $\qquad i_{st}(t) = A \cdot \sin(\omega t - \alpha).$
Both the current and the voltage are sinusoidal in shape with the same frequencies. But they differ in phase. "The current lags the voltage".

- In the special case $L = 0$, current and voltage are in phase, i.e. $\alpha = 0$.
- In the special case $R = 0$, we have $\alpha = \pi/2$.

3.14 Catenary

The aim is to calculate the shape of a chain or cable sagging due to its own weight. Regardless of whether we are dealing with a necklace, a power line, the cable of a cable car, a chain or rope barrier, the derivation remains valid as long as the object is very flexible and has a uniformly distributed mass.

Let the curve of the chain, called the **catenary**, be described by the function $y = f(x)$. The figure shows a small portion of the sagging cable between x and $x + \Delta x$. The forces on the left and right act at the ends tangentially and the weight ΔG acts vertically.

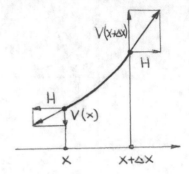

Newton's law of inertia states that the vector sum of all the three forces equals $\vec{0}$. Hence the horizontal component is

$$H(x) = H = \text{ constant.}$$

The vertical component $V(x + \Delta x)$ on the right exceeds the one on the left by ΔG:

$$V(x + \Delta x) - V(x) = \Delta G. \tag{3.11}$$

The length $\Delta \ell$ is computed by the integral

$$\Delta \ell = \int_{x}^{x+\Delta x} \sqrt{1 + f'(u)^2} \cdot du.$$

By virtue of the Mean Value Theorem for integrals there is an intermediate point $z \in (x, x + \Delta x)$ such that

$$\Delta \ell = \int_{x}^{x+\Delta x} \sqrt{1 + f'(u)^2} du = \sqrt{1 + f'(z)^2} \cdot \Delta x. \tag{3.12}$$

Furthermore,
$$\Delta G = \rho \cdot \Delta \ell \cdot g, \tag{3.13}$$

in which g is the gravitational acceleration, ρ the density per unit length with dimension $[\rho] = \text{kg/m}$.
Substituting (3.12) into (3.13) and the result into (3.11) yields

$$V(x+\Delta x) - V(x) = \rho \cdot g \cdot \sqrt{1 + f'(z)^2} \cdot \Delta x.$$

Division by Δx leads to

$$\frac{V(x+\Delta x) - V(x)}{\Delta x} = \rho \cdot g \cdot \sqrt{1 + f'(z)^2}.$$

Taking the limit as $\Delta x \to 0$ implies $z \to x$ and hence

$$V'(x) = \rho \cdot g \cdot \sqrt{1 + f'(x)^2}. \tag{3.14}$$

As the force has a tangential direction,

$$f'(x) = \frac{V(x)}{H} \implies V(x) = H \cdot f'(x) \implies V'(x) = H \cdot f''(x).$$

Substitution into (3.14) implies

$$f''(x) = \frac{\rho g}{H} \cdot \sqrt{1 + f'(x)^2} = k \cdot \sqrt{1 + f'(x)^2} \qquad \text{where } k = \frac{\rho g}{H}.$$

On first sight, this is a second order differential equation. But if we set $g(x) = f'(x)$, then we obtain the following autonomous first order differential equation in the unknown function $g(x)$:

$$g' = k \cdot \sqrt{1 + g^2}.$$

The Mathematica command $\texttt{DSolve[g'[t] == k}\sqrt{1+g[t]^2}\texttt{,g[t],t]}$ produces the general symbolic solution $g(x) = \sinh(kx + C)$.
Integration yields the equation of a catenary with two parameters C and D:

$$f(x) = \frac{1}{k} \cosh(kx + C) + D.$$

The dimensions are consistent:

$$[k] = \frac{\text{kg/m} \cdot \text{m/s}^2}{\text{kg} \cdot \text{m/s}^2} = \frac{1}{\text{m}}.$$

This implies that the shape of the catenary does not depend on the weight. It is fully determined by the two suspension points of the chain ends $A(x_1, y_1)$, $B(x_2, y_2)$, and the length L of the chain. Now we wish to compute the parameters k, C, D fitting the given suspension points and the given length.

$$L = \int_{x_1}^{x_2} \sqrt{1 + [f'(x)]^2} \cdot dx = \int_{x_1}^{x_2} \sqrt{1 + \sinh^2(kx + C)} \cdot dx$$

$$= \int_{x_1}^{x_2} \cosh(kx + C) \cdot dx = \frac{1}{k} \left(\sinh(kx_2 + C) - \sinh(kx_1 + C) \right).$$

The following calculation is based on a suggestion in [18] with the following abbreviations for $i = 1, 2$:

$$s_i = \sinh(kx_i + C), \quad c_i = \cosh(kx_i + C), \quad y_i = f(x_i) = \frac{1}{k} c_i + D.$$

Hence

$$L = \frac{1}{k}(s_2 - s_1) \quad \text{and} \quad y_1 - y_2 = \frac{1}{k}(c_1 - c_2).$$

Now we use the two identities

$$\cosh^2 z - \sinh^2 z = 1 \quad \text{and} \quad \cosh u \cosh v - \sinh u \sinh v = \cosh(u - v),$$

to compute the term

$$\sqrt{L^2 - (y_1 - y_2)^2} = \frac{1}{k}\sqrt{s_2^2 + s_1^2 - 2s_1 s_2 - c_2^2 - c_1^2 + 2c_1 c_2}$$

$$= \frac{1}{k}\sqrt{2\cosh[k(x_1 - x_2)] - 2}.$$

With the identity $\sqrt{\cosh z - 1} = \sqrt{2}\sinh\frac{z}{2}$, we have

$$\sqrt{L^2 - (y_1 - y_2)^2} = \frac{2}{k}\sinh\left[\frac{k(x_1 - x_2)}{2}\right].$$

Dividing by $x_1 - x_2$ yields

$$\frac{\sqrt{L^2 - (y_1 - y_2)^2}}{x_1 - x_2} = \frac{\sinh\left[\frac{k(x_1 - x_2)}{2}\right]}{\frac{k(x_1 - x_2)}{2}}.$$

With $\alpha = \dfrac{k(x_1 - x_2)}{2}$, we obtain a nonlinear equation

$$\frac{\sinh\alpha}{\alpha} = \frac{\sqrt{L^2 - (y_1 - y_2)^2}}{x_1 - x_2} \tag{3.15}$$

with exactly one positive solution α.

Algorithm for Calculating the Parameters k, C, D:

(a) Solve Equation (3.15) numerically for α. Then $k = \dfrac{2\alpha}{x_1 - x_2}$.

(b) Solve the nonlinear equation

$$y_1 - y_2 = \frac{1}{k}\left[\cosh(kx_1 + C) - \cosh(kx_2 + C)\right]$$

numerically for the parameter C.

(c) Finally compute $D = y_1 - \frac{1}{k}\cosh(kx_1 + C)$.

3.15 Global Warming

A simple model for ice formation on the surface of water based on the work of Josef
Stefan (1835–1893) [7] is described in [21]. During ice formation, when water freezes
to ice, heat is generated at the ice bottom. The model is based on the assumption that
this heat is conducted through the layer of ice with a constant temperature gradient.
This means that we have a linear temperature drop from bottom to top. Of pressing
importance at the present time is the melting of ice, for which the model also works.
In this case, we have a linear temperature rise from bottom to top. In addition, it is
assumed that there is no internal heat source and there is no heat transport from sea
to ice. In this setting, ice formation and ice melting obeys the law

$$\rho L \frac{dH}{dt} = \kappa [T_f - T_0(t)] \cdot \frac{1}{H},$$

in which $H(t)$ is the thickness of the ice layer as a function of time, $\rho = 0.9$ g/cm^3
the densitiy of ice, $\kappa \approx 2.2$ W/(K$\cdot$m) the heat conductivity of snow free ice, $T_f =
-2$ °C the freezing temperature of salt water (0 °C is that of fresh water), $T_0(t)$ the
temperature on the surface as a function of time, and $L = 335$ J/g the heat of fusion
(heat of solidification) for ice. We solve the above differential equation for H by
separating variables:

$$\int H \cdot dH = \frac{\kappa}{\rho L} \int_0^t [T_f - T_0(\tau)] \cdot d\tau + C.$$

Substituting $a^2 = \frac{2\kappa}{\rho L}$ and imposing the initial condition $H(0) = H_0$ yields

$$H^2 = a^2 \int_0^t [T_f - T_0(\tau)] \cdot d\tau + H_0^2 \implies H(t) = \sqrt{H_0^2 + a^2 \int_0^t [T_f - T_0(\tau)] d\tau}.$$

In this context, the unit of time is d (days). With the given data for snow free ice

$$a^2 = \frac{2 \cdot 2.2 \text{ W}}{\text{K} \cdot \text{m} \cdot 0.9 \text{ g/cm}^3 \cdot 335 \text{ J/g}} = \frac{4.4 \text{ cm}^3}{\text{K} \cdot 100 \text{ cm} \cdot 0.9 \cdot 335 \text{ s}} = 12.61 \frac{\text{cm}^2}{\text{K} \cdot \text{d}},$$

because in terms of dimensions $\frac{J}{W} = s$ and $1 \text{ d} = 24 \cdot 3600$ s.
For a moment let us assume that the temperature difference $T_f - T_0(t)$, having a
period of 1 year, has the mean value 0 (no global warming). Then the ice thickness
varies periodically with a period of 365 d and its average remains constant.

Next we postulate that global warming has taken place and that the air temperature
T_0 has risen by $T = 1.5$ °C. Then the temperature difference still has an annual period, but its mean value now is -1.5 °C. Furthermore, it is observed to vary between
23.5 °C and -26.5 °C. With the assumed periodic function

$$T_f - T_0(t) = 25 \cdot \cos(\omega t) - 1.5, \quad \text{where} \quad \omega = \frac{2\pi}{365},$$

integration yields

$$H(t) = \sqrt{H_0^2 + a^2 \left(\frac{25}{\omega} \sin(\omega t) - 1.5t \right)}.$$

---

[7] Josef Stefan was a mathematician and physicist of Slovenian mother tongue. His most significant
scientific contribution was the Stefan–Boltzmann Law of Radiation.

Here is its graph, where $H_0 = 300$ cm, together with the one without the sine term (average values):

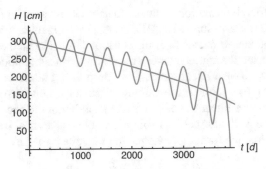

As can be read from the graph, the model predicts that the ice layer would disappear within 11 years.

More generally, let us consider **any function** $T_0(t)$ **that has an annual periodicity** with a constant mean temperature rise (global warming) of T °C against T_f. Thus the average air temperature is $T_f + T$. The mean integrated function

$$\overline{H}(t) = \sqrt{H_0^2 - a^2 \cdot T \cdot t}$$

allows us to get a useful upper bound for the time t_0 at which the water turns ice free, namely the zero of $\overline{H}(t)$:

$$t_0 = \frac{H_0^2}{a^2 \cdot T} \quad \text{where } [t_0] = \text{d}, \quad [H_0] = \text{cm} \quad \text{and} \quad 12.61 \, \frac{\text{cm}^2}{\text{K} \cdot \text{d}}.$$

The following graph of $t_0(T)$ visualizes the case when $H_0 = 300$ cm for the range of mean temperature rises $0.25 \, °\text{C} \leq T \leq 3 \, °\text{C}$ that are assumed constant:

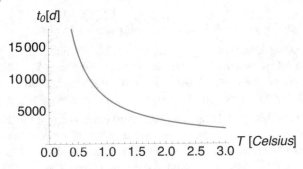

For $T = 1.5 \, °\text{C}$ the maximum duration would be $t_0 = 4760$ d ≈ 13 a until the ice layer disappears, regardless of the shape of the annual function!

Ice layer of a freshwater lake: Let us inquire into how the ice thickness increases in winter on a freshwater lake with a daily constant mean air temperature of $T_0 < 0 \, °\text{C}$.

For $H_0 = 0$ we obtain $\overline{H}(t) = a \cdot \sqrt{|T_0| \cdot t}$.

Example: After 50 days, $T_0 = -10 \, °\text{C}$ implies an ice thickness of about $\overline{H}(50) = \sqrt{12.61} \cdot \sqrt{500} = 79$ cm.

3.16 Brachistochrone Problem

This is a historically significant problem that was solved by Johann Bernoulli (1667–1748). In 1696 it prompted Jakob I. Bernoulli (1655–1705) to develop the Calculus of Variations.

A point mass with initial velocity $v = 0$ slides from a given initial point A to a given terminal point B in the absence of friction and under the influence of gravitation. Find the path of **fastest descent**.

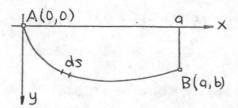

The problem may also be worded as follows: Which is the path a person must take when slalom skiing on an inclined plane in order to get from start A to finish B in the shortest time, assuming that there is no friction (which is practically almost true).

Idea: We consider thin horizontal layers, see the following figure. In each layer the velocity is practically constant, but increasingly faster downwards.

Fermat's Principle says that for the **fastest** path Snell's Law of Refraction holds:

$$\frac{\sin\alpha_1}{\sin\alpha_2} = \frac{v_1}{v_2}.$$

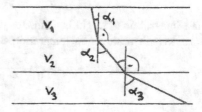

This is valid here just as it is in optics for a ray of light passing through two optic media in which light travels with different velocities.

Snell's law can be rewritten as follows:

$$\frac{\sin\alpha_1}{v_1} = \frac{\sin\alpha_2}{v_2}.$$

Repeating this reasoning for each of the following layers implies

$$\frac{\sin\alpha_1}{v_1} = \frac{\sin\alpha_2}{v_2} = \frac{\sin\alpha_3}{v_3} = \ldots = \text{constant}.$$

If the number of layers tends to infinity and the thickness of layers tends to 0, we obtain

$$\frac{\sin\alpha}{v} = \text{constant} \qquad (3.16)$$

where v and α vary continuously.

Let us now express v and α in terms of the height y and the slope y':
For the height y we use the law of energy conservation:

$$\frac{m}{2}v^2 = mgy \Longrightarrow v = \sqrt{2gy}.$$

By inspecting the infinitesimal triangle, we find

$$\sin\alpha = \frac{dx}{\sqrt{dx^2 + dy^2}} = \frac{1}{\sqrt{1 + (\frac{dy}{dx})^2}}.$$

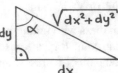

Substituting the two terms into (3.16) yields

$$\frac{1}{\sqrt{1+(y')^2}} \cdot \frac{1}{\sqrt{2gy}} = \text{constant}.$$

Simpler still (with the positive constant on the right):

$$[1+(y')^2] \cdot y = k^2.$$

Solving for y' yields a separable differential equation[8]

$$\frac{dy}{dx} = \sqrt{\frac{k^2 - y}{y}}.$$

Integration yields
$$\int dx = \int \sqrt{\frac{y}{k^2 - y}}\, dy + C. \qquad (3.17)$$

The integral on the right is elementary, but it is not explicitly solvable for y. Therefore, we substitute a new integration variable u:

$$y(u) = k^2 \sin^2 \frac{u}{2} \Longrightarrow \sqrt{\frac{y}{k^2 - y}} = \sqrt{\frac{k^2 \sin^2 \frac{u}{2}}{k^2 (1 - \sin^2 \frac{u}{2})}} = \frac{\sin \frac{u}{2}}{\cos \frac{u}{2}}. \qquad (3.18)$$

Hence

$$\frac{dy}{du} = k^2 \sin\frac{u}{2}\cos\frac{u}{2} \Longrightarrow dy = k^2 \sin\frac{u}{2}\cos\frac{u}{2} \cdot du. \qquad (3.19)$$

Substituting (3.18) and (3.19) into (3.17) yields

$$\int dx = k^2 \int \sin^2 \frac{u}{2} \cdot du + C.$$

Using the trigonometric identity $\sin^2 \frac{u}{2} = \frac{1}{2}(1 - \cos u)$, we obtain x and y as functions of u:

$$x(u) = \frac{k^2}{2}(u - \sin u) + C \qquad y(u) = \frac{k^2}{2}(1 - \cos u).$$

---

[8] This derivation corresponds to the historical solution of Johann Bernoulli. In [9] the differential equation was deduced by Calculus of Variations, which in the mid-18th century was substantially enhanced by Leonhard Euler (1707–1783) and Joseph-Louis Lagrange (1736–1813).

As the starting point A for $t = 0$ is at the origin, the integration constant $C = 0$.
Substituting $t = u$ and $a = \frac{k^2}{2} > 0$ yields the following parametrized curve:

$$\vec{r}(t) = \begin{pmatrix} x(t) \\ y(t) \end{pmatrix} = a \begin{pmatrix} t - \sin t \\ 1 - \cos t \end{pmatrix}. \tag{3.20}$$

The answer is the same as in (1.12). Hence it is an **arc of the cycloid**, traced out
by a point on the rim of a rolling wheel of radius a. Obviously, the figure obtained
earlier must be reflected in the x-axis.

If the terminal point B is only slightly lower than the initial point A, then the curve of
fastest descent looks like the following figure. The radius a is uniquely determined
by the positions of the points A and B.

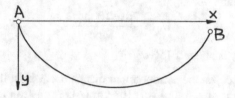

3.17 Cosmology

3.17.1 Historical Facts

The Belgian theologian and astrophysicist Abbé Georges Lemaître (1894–1966) and
the Russian meteorologist and mathematician Aleksandrovich Friedmann (1888–
1925) independently developed a relativistic model for describing the extension of
the universe over time.

Lemaître could convince Albert Einstein (1879–1955) of his model in person. Einstein had previously believed in a static universe. Lemaître was probably the first in
proving that the universe was expanding. He published his ideas two years prior to
Edwin Hubble (1889–1953), the American astronomer. The red shifts measured by
the Hubble Telescope confirmed the expansion of the cosmos. Cosmological models
are based on Einstein's assumption that the universe is homogeneous and isotropic
on a grand scale (for distances that are at least 300 light years), i.e. the universe
viewed from each point and in each direction is the same. On the other hand it
allows the possibility of expansion and contraction.

Lemaître is considered the father of the big bang theory and was therefore honored
by Pope Pius XII.

3.17.2 Extent and Age of the Universe

Once it was clear that the universe was expanding, one successfully found a quantitative linear relationship between the escape velocity v and the distance D of galaxies in local universes (not valid for all galaxies, but fairly well for most of them):

$$v = H_0 \cdot D. \tag{3.21}$$

By local we mean that the law is valid only for distances that are small compared to the global extent of the universe.

So, with increasing distance from local galaxies the escape velocity increases linearly. The Hubble constant H_0, significant in cosmology, was determined from numerous measurements:

$$H_0 \approx \frac{70 \pm 10 \text{ km/s}}{\text{Mpc}}.$$

Thus its uncertainty is about $\pm 15\%$.

The unit 1 pc (parsec) is the astronomical distance at which the distance between the earth and the sun of 1 AU $= 1.5 \cdot 10^8$ km (AU for astronomical unit) subtends an angle of one arcsecond.

Another astronomical unit of length is the distance traveled by the light in one year. Hence

- 1 ly $\approx 10^{13}$ km (ly for light year),

- 1 pc $= 3.1 \cdot 10^{13}$ km $= 3.1$ ly $= 2 \cdot 10^5$ AU,

- $1 Mpc = 3.1 \cdot 10^{19}$ km $= 3.1 \cdot 10^6$ ly $= 2 \cdot 10^{11}$ AU.

Interestingly, the reciprocal of H_0 (where $1 \text{ a} = 3.15 \cdot 10^7$ s) provides the age of the universe:

$$\frac{1}{H_0} \approx \frac{1}{2.3 \cdot 10^{-18} \text{ s}^{-1}} \approx 0.43 \cdot 10^{18} \text{ s} = 0.136 \cdot 10^{11} \text{ a} = 13.6 \text{ billion years.}$$

The scale of the universe is hard to grasp for common people. For instance, the distance D to the Andromeda Galaxy, the nearest galaxy that is visible to the naked eye, is $D = 2.5$ ly ≈ 0.8 Mpc. Hence according to (3.21), the Andromeda Galaxy moves away from us at a radial velocity of $v \approx 56$ km/s. Our Milky Way and the Andromeda Galaxy are spiral in shape, both having a diameter of about $100,000$ light years.

The impressively general theory of so-called **symmetry breaking and bifurcation**[9] explains the emergence of the various types of galaxies. To the nonexperts, we recommend the excellently written popular science book [14], in which galaxies are explored as well as applications from various fields of physics.

---

[9] Bifurcation is dealt with in [2] Section 2.5 and bifurcation in the context of Symmetry Breaking is dealt with in [11] Section 6.1.

"Die Welt" reported in March 2016:

"One can hardly get closer to the big bang. The Hubble telescope discovered a galaxy from the beginnings of the universe. Its light shines to us from a record distance of 13.4 billion light years. We see the galaxy from a time when the universe was only three percent of its current age. In the past, astronomers could only provide estimates of the galaxy's distance. It is only now that the Hubble telescope allows an accurate measurement."

Also in the year 2016, based on shots taken by the Hubble telescope, astronomers reported that the number of galaxies in the universe had been underestimated by ten times. Before the year 2000, the number was estimated to be 100 billion. Among the now newly estimated 1000 billion about 90% are invisible. Furthermore, the estimated number of stars in the Milky Way alone is 100 billion!

3.17.3 Olbers' Paradox

Under the assumption that the universe in Euclid's geometry is infinitely large and static, we shall conclude that the night sky should be bright. But this completely contradicts our observation, implying that the assumption is false.

Let n be the number of stars per unit volume, considered constant in space and time, and s the average radius of the stars, considered spherical. Such a star at a distance R from the earth subtends a solid angle of $\frac{\pi s^2}{R^2}$.

In a hollow sphere with radii R and $R + dR$ there are $n \cdot dV = dR \cdot 4\pi R^2 n$ stars. Together they subtend an infinitesimal solid angle

$$d\omega = \frac{\pi s^2}{R^2} \cdot dR \cdot 4\pi R^2 n = 4\pi^2 n s^2 \cdot dR, \quad \text{which is independent of } R.$$

The integral over all hollow spheres for the solid angle diverges:

$$\omega = 4\pi^2 n s^2 \int_0^\infty dR = \infty.$$

Hence there are overlappings of the circular disks. Due to homogeneity and isotropy, all of the sky should be illuminated.

3.17.4 Friedmann-Lemaître Equation

"Derivation" according to [33]: Let us begin with Newton's mechanics, even though the General Relativity Theory (GRT) must certainly be involved. However, as will be seen, the GRT will come into play through a surprisingly simple artifice.

For the sake of simplicity, we consider a spherical universe. However, it should be noted that the result holds in general.

Let r_0 be the radius of the universe and ρ_0 the density of the masses at the present time t_0. The radius as a function of time is

$$r(t) = a(t) \cdot r_0, \tag{3.22}$$

where the scale factor a is dimensionless.

Now we assume that the total mass of the universe is constant.[10] Hence

$$\rho(t)\frac{4\pi}{3}r(t)^3 = \rho_0 \frac{4\pi}{3}r_0^3.$$

Substituting Equation (3.22) and canceling common factors yields $\rho(t) = \dfrac{\rho_0}{a(t)^3}$.

Because of the homogeneity of the universe, we may assume that the total mass $M = \rho_0 \cdot \frac{4\pi}{3}r_0^3$ of the universe is concentrated at its center. Then by Newton's law of motion, the gravitational force acting on point mass m at distance r from the center of the universe is

$$m \cdot \ddot{r} = -\frac{GmM}{r^2} = -\frac{Gm}{r^2}\frac{4\pi}{3} \cdot r_0^3 \cdot \rho_0.$$

Substituting Equation (3.22) with $\ddot{r} = \ddot{a} \cdot r_0$, and dividing by m yields

$$\ddot{a} = -\frac{4\pi G\rho_0}{3} \cdot \frac{1}{a^2}.$$

We multiply both sides by $2\dot{a}$ to obtain

$$2\dot{a}\ddot{a} = -\frac{8\pi G\rho_0}{3} \cdot \frac{\dot{a}}{a^2},$$

which may be rewritten as

$$\frac{d}{dt}\dot{a}^2 = \frac{8\pi G\rho_0}{3} \cdot \frac{d}{dt}a^{-1}.$$

Integrating both sides with respect to t yields the **Friedmann-Lemaître Equation**

$$\dot{a}^2 = \frac{8\pi G\rho_0}{3} \cdot \frac{1}{a} - Kc^2. \tag{3.23}$$

The artifice announced earlier enters here: We interpret the integration constant as Kc^2, where c is the velocity of light and K is the curvature of the four-dimensional space-time universe (dimension $[K] = m^{-2}$). Thus (3.23) is the relativistic cosmological model.

Remark: Einstein added on the right-hand side his controversial cosmological constant $\frac{\Lambda}{3}$ as a summand, which he interpreted in a contradictory fashion. In fact, he later referred to it as his "biggest blunder". So, we refrain from it.

---

[10] According to the GRT, this assumption is incorrect, because the equation $E = mc^2$ says that mass can be converted into radiation energy and vice versa.

The Gaussian curvature K of a point on a 2-dimensional surface is the product of the principal curvatures k_1 and k_2 of two normal sections of the surface at that point.

If $K > 0$, then the point is elliptic.
If $K < 0$, then the point is hyperbolic.
For a sphere with radius R, we have $K = 1/R^2 = $ constant.
Examples for $K = 0 = $ constant are the cylinder, the cone, and the plane.

3.17.5 Einstein-De Sitter Model

In this model[11] we assume that $K = 0$. As the universe is presently expanding, $\dot{a} > 0$. Equation (3.23) leads to a separable differential equation for the scale factor a:

$$\dot{a} = \sqrt{\frac{8\pi G\rho_0}{3} \cdot \frac{1}{a}} = \sqrt{\frac{\lambda}{a}} \quad \text{where} \quad \lambda = \frac{8\pi G\rho_0}{3}.$$

$$\int \sqrt{a} \cdot da = \int \sqrt{\lambda} \cdot dt \implies \frac{2}{3}a^{3/2} = \sqrt{\lambda} \cdot t + C \implies a = (\frac{3}{2}\sqrt{\lambda}t + C)^{2/3}.$$

Choosing $C = 0$ yields

$$a(t) = kt^{2/3} \quad \text{where} \quad k = (\frac{3\sqrt{\lambda}}{2})^{2/3}.$$

This model postulates that from the big bang ($a = 0$) at time $t = 0$ onward, the universe keeps expanding forever.
Moreover, due to $\dot{a}(0) = \infty$, right after the big bang the universe expanded at racing speed.

3.17.6 Model for a Universe with Positive Space Curvature

$$\dot{a} = \sqrt{\frac{\lambda}{a} - K \cdot c^2} = \sqrt{Kc^2\left(\frac{\lambda}{a \cdot Kc^2} - 1\right)} = \sqrt{K}c \cdot \sqrt{\frac{\lambda}{aKc^2} - 1}.$$

Substituting the constant $\mu = \dfrac{\lambda}{Kc^2} = \dfrac{8\pi G\rho_0}{3Kc^2}$ yields the separable differential equation

$$\dot{a} = \sqrt{K}c \cdot \sqrt{\frac{\mu}{a} - 1} = \sqrt{K}c\sqrt{\frac{\mu - a}{a}}.$$

$$\sqrt{K}c \int dt = \int \sqrt{\frac{a}{\mu - a}} \cdot da + C.$$

---

[11] Willem De Sitter (1872–1934) was a Dutch astronomer. He showed that the speed of light was independent of the speed of the light source, thus confirming the special theory of relativity.

This integration problem is identical to (3.17), except for the factor $\sqrt{Kc}$! But be careful with the variable names: x corresponds to t, y corresponds to a, and the former parameter t will now be called s.

We may apply the calculation method by substituting $a(s) = \mu \sin^2 \frac{s}{2}$ in the right-hand integral to obtain

$$\sqrt{Kc} \cdot t(s) = \frac{\mu}{2}(s - \sin s) + C.$$

Choosing $C = 0$ and $A = \frac{\mu}{2}$ yields an **affinely distorted arc of a cycloid** that is parametrized as follows:

$$t(s) = \frac{A}{\sqrt{Kc}} \cdot (s - \sin s),$$

$$a(s) = A(1 - \cos s).$$

Here the parameter range is $0 \le s \le 2\pi$. Thus the curve starts for $s = 0$ at the origin of the (t, a)-coordinate system.

When checking the dimensions, make sure that the curvature has the dimension m^{-2}. The quantity A indeed is dimensionless and $\frac{A}{\sqrt{Kc}}$ has the dimension of time.

If this model is used, then the origin corresponds to the big bang $(a = 0)$ at time $t = 0$. The model predicts that the universe expands until time $t(\pi) = T = \frac{\pi A}{\sqrt{Kc}}$ with maximum $a(\pi) = 2A$, then shrinks and finally collapses at time $t = 2T$.

It is remarkable that the tangent of $a(t)$ is vertical at the times of big bang and collapse. This means that the expansion factor a for $t = 0$ and $t = 2T$ changes infinitely fast. Compare with Exercise 11 where the tangent of the graph $D(h)$ is also vertical at $h = 0$.

3.18 Orthogonal Trajectories

In certain physical applications, it is necessary to determine the orthogonal trajectories of a given family of curves. Orthogonal trajectories are curves that intersect the curves of the given family perpendicularly.

Let the given family of curves be described by the solutions of the differential equation
$$y' = f(x, y).$$

Then the orthogonal trajectories are described by the solutions of the differential equation
$$y' = -\frac{1}{f(x, y)},$$

because the product of the two slopes must equal -1.

Remark: One must clearly distinguish between the two families of curves, even though they are both denoted by y.

Example 3.28

Let the given family of curves be described by the equation $x \cdot y(x) = k$.

For

(a) $k > 0$, the curves are hyperbolas in the first and third quadrants with axis of symmetry $y = x$,

(b) $k < 0$, the curves are hyperbolas in the second and fourth quadrants with axis of symmetry $y = -x$,

(c) $k = 0$, the curves are the two coordinate axes. They are also the asymptotes of the hyperbolas in (a) and (b).

By differentiating the given equation, we obtain the differential equation for the family of curves:

$$y + x \cdot y' = 0 \Longrightarrow y' = -\frac{y}{x}. \tag{3.24}$$

Therefore, the separable differential equation for the orthogonal trajectories is

$$y' = \frac{x}{y} \Longrightarrow \int y \cdot dy = \int x \cdot dx + C \Longrightarrow y^2 = x^2 + C.$$

For

(a) $C = a^2 > 0$, the equation $\frac{y^2}{a^2} - \frac{x^2}{a^2} = 1$ describes equilateral hyperbolas opening up and down,

(b) $C = -a^2 < 0$, the equation $\frac{x^2}{a^2} - \frac{y^2}{a^2} = 1$ describes equilateral hyperbolas opening left and right,

(c) $C = 0$, the two solutions are $y = \pm x$. They are also the asymptotes of the hyperbolas in (a) and (b).

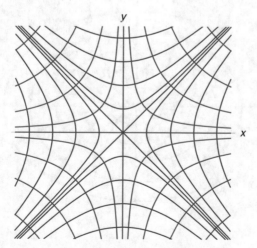

The figure depicts the family of orthogonal trajectories (red curves), which is obtained by just rotating the given family (blue curves) through $45°$.   ◇

Example 3.29 Let us compute the orthogonal trajectories of the family of parabolas

$$y(x) = kx^2.$$

Differentiation gives $y' = 2kx$.

Substituting the parameter $k = \dfrac{y}{x^2}$ yields a differential equation for the family of parabolas:

$$y' = \frac{2y}{x}.$$

This leads to a separable differential equation for the orthogonal trajectories:

$$y' = -\frac{x}{2y}.$$

Hence the solutions satisfy $\qquad y^2 + \dfrac{x^2}{2} = C.$

$C = a^2$ implies

$$\frac{x^2}{(\sqrt{2}a)^2} + \frac{y^2}{a^2} = 1,$$

describing a family of ellipses whose semiaxes have ratio $\sqrt{2} : 1$.

The figure shows members of the family of parabolas (red) and members of the corresponding family of ellipses (blue) as orthogonal trajectories.

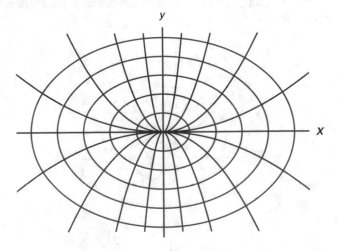

3.19 Exercises Chapter 3

49. Coffee. A cup of coffee has an initial temperature of 95 °C, one minute later it is 86 °C. When has the temperature dropped to 37 °C, if the room temperature is 20 °C? Sketch a graph.

50. Radium. The half-life of Ra 226 is $T_{1/2} = 1623$ years.
How long does it take for 10% of the radioactivity to dissipate?

51. Compound Interest. A principal of initially 1 million euros is invested at $p\%$ interest.

(a) What is the principal after one year and after n years, if the interest is compounded continuously?

(b) What is the principal after one year and after n years, if the interest is compounded annually?

(c) Compare the two values for $p = 3\%$ and $n = 10$ years.

(d) Compare the two values for a negative interest rate of $p = -3\%$ after $n = 10$ years.

52. Barometric Formula for the Density.
(a) Explain why the density $\rho(h)$ in isothermal atmosphere is

$$\rho(h) = \rho_0 \cdot e^{kh},$$

with the same k-value as in the barometric formula for the pressure $p(h)$.
Remark: The density $\rho_0 \approx 1.225 \mathrm{kg/m^3}$ holds for temperature $T(0) = 15$ °C at sea level.
(b) Compare density and pressure at altitude 80 km with those at sea level for $T(0) = 15$ °C = const and $k = 0.00011861$ from Section 3.11.1.
(c) Show that for equidistant function values $T_1, T_2, \ldots, T_n$ on the interval $[0, h]$ the

(discrete) harmonic mean T_m defined as $\dfrac{1}{T_m} = \dfrac{1}{n} \displaystyle\sum_{i=1}^{n} \dfrac{1}{T_i}$ tends to the (continuous)

harmonic mean $\dfrac{1}{T_m} = \dfrac{1}{n} \displaystyle\int_0^h \dfrac{1}{T(s)} ds$ as $n \to \infty$.

53. Chernobyl and Vegetables. The radioactive isotope iodine 131 has a half-life of $T_{1/2} = 8$ days. To what percentage did its radioactivity drop after two months? Were the vegetables contaminated by iodine 131 safe for consumption two months after the nuclear accident?

54. Growth of a Cell. The volume V of a cell grows proportionally to its surface area. Assuming that the cell is spherical in shape, show that $\dot{V} = k \cdot V^{2/3}$.
Determine the growing radius $r(t)$ as a function of time.

55. Mixing Problem. A tank originally contains 120 l of pure water. A brine containing 50 g salt/l flows at a rate of 2 l/min into the tank, and at the same time 2 l/min of the well stirred solution flows out of a hole. Let $S(t)$ be the amount of salt at time t.

(a) Determine first the limiting amount S_∞, and then the differential equation.

(b) When is $S = 5.90$ kg, and when is it 5.99 kg? Compare.

56. Conical Tank. An empty open conical tank of radius R and height H is filled through an inlet from time $t = 0$ onward at a rate of L liters/second.

(a) How much time t_E passes until the container is filled?

(b) Compute and sketch the level $h(t)$ of the liquid and its velocity $v(t)$ as functions of time t.

(c) Check your answer for $h(t)$ by means of your answer to (a) and, in particular, discuss $v(0)$.

(d) Compute the velocity $w(h)$ as a function of h. Also compute $\frac{w(H)}{w(H/2)}$. Verify that $w(H) = v(t_E)$.

(e) Compute numerically t_E, $v(t_E)$, $v(t_E/2)$, $h(t_E/2)$ for $R = H = 1$ m, $L = 2$ l/s.

57. Smoking Problem. A room measuring 4 m $\times$ 5 m $\times$ 2.5 m is originally free from the toxic carbonmonoxide CO. At time $t = 0$, someone starts smoking. Smoke is expelled into the air at 3 l of smoke/min. The smoke contains 4 volume percent of CO. Per minute 3 l of the well-mixed air-and-smoke mixture leave the room. Let $V(t)$ be the volume of CO at time t.

(a) Find a differential equation.

(b) If the percentage of volume of CO reaches 0.0012%, it will be harmful to human health. When does this happen?

58. Motor Boat. The total mass m of two people including their small motor boat is 500 kg. The motor generates a constant force of $F = 200$ N. Let the resistance force of water be $W = k \cdot v$ where $k = 10\frac{N}{m/s}$.

(a) By use of Newton's law   *mass $\times$ acceleration = resultant force*   verify that the differential equation for the speed $v(t)$ is

$$m\dot{v} = F - kv.$$

(b) Sketch the direction field of the autonomous differential equation and compute $v_\infty = v(\infty)$.

(c) At the start $v_0 = 0$. How long does it take for the speed to reach $0.9 \cdot v_\infty$? What distance did the boat travel by that time?

59. Ball in a Gravitational Funnel. Analogously to Section 3.9 compute the geometrical shape $r(h)$ of a gravitational funnel for a rotating ball of radius ρ, mass m, moment of inertia $J = \frac{2}{5}m\rho^2$, and show that the angular frequency is

$$\Omega(h) = \frac{1}{R \cdot v_0^{7/5}} \cdot v(h)^{-2/5}.$$

Compute the numerical value of the radius at the lower opening $r(0.4$ m$)$ for $R = r(0) = 0.2$ m, $v_0 = 1$ m/s, and compare your answer with that of Section 3.9.

60. Leaking Container. A cylindrical barrel of diameter 1 m and height 1 m is completely filled with water. At the bottom there is a hole of radius 1 mm. How long does it take to drain the barrel completely?

61. Free Fall inside a Liquid. A container is filled with a liquid of viscosity η. At time $t = 0$ a small gold ball of radius $r = 3$ mm and density $\rho = 19.3$ g/cm³ is released on the liquid's surface ($v_0 = 0$).
If buoyancy is neglected (as the ball is small), then the time-varying velocity $v(t)$ satisfies

$$\dot{v}(t) + k \cdot v(t) = g$$

where $k = \frac{6 \cdot \pi \cdot \eta \cdot r}{m}$, m is the mass of the gold ball, and g is the acceleration of gravity.

(a) Express the limiting velocity v_∞ in terms of g and k, without solving the differential equation.
(b) For given k and g, compute symbolically the velocity function $v(t)$ and the level of sinking $y(t)$.
(c) Compute numerically the limiting velocities v_∞ of the gold ball in glycerine ($\eta = 15.0$ kg·m⁻¹·s⁻¹) and in olive oil ($\eta = 1.07$ kg·m⁻¹·s⁻¹). Pay attention to the dimensions.
(d) Consider a steel ball ($\rho = 7.8$ g/cm³) of the same size ($r = 3$ mm) in glycerine. How does the limiting velocity of the steel ball compare to that of the gold ball? (Find the ratio of the two limiting velocities.)

62. Suspension Bridge. Compute the curve $y = f(x)$ of the supporting cables of a suspension bridge. In the model take into account the weight of the roadway, but neglect the weight of the suspension cables. Let ρ be the constant mass of the roadway per running meter. As in the problem of the catenary, first establish a differential equation for $f(x)$.

63. Specified Elasticity Function. Compute the function f, given its constant elasticity function $\varepsilon_{f,x} = a$.

64. Function = Elasticity Function. Find the functions $f(x)$ that are identical to their elasticity functions $\varepsilon_{f,x}$ and sketch their graphs.

65. Tree Growth According to Chapman-Richards. We wish to find the height $h(t)$ of a tree as a function of time t.
Both the increase dm/dt of its mass m per unit of time and the loss of water per unit of time are proportional to h^3.
Let the absorbed heat output of solar radiation be proportional to $h^{2.5}$. Reason: Photosynthesis works not only on the leaves of the canopy (where it would be proportional to h^2), but also, albeit to a decreasing degree, on the leaves of the interior of the tree (where it would be proportional to h^3), hence its proportionality to $h^{2.5}$.
The power balance for $m(t) = a \cdot h(t)^3$ is given by

$$\frac{dm}{dt} = b \cdot h^{2.5} - c \cdot h^3$$

with given positive parameters a, b, c, which are characterized by the type of tree.

(a) Show that the following differential equation results from the power balance:

$$\dot{h} = \frac{b}{3a} \cdot \sqrt{h} - \frac{c}{3a} \cdot h.$$

(b) Use a suitable substitution to convert the problem into a linear differential equation in u and then determine the growth function $h(t)$ for the initial value $h(0) = h_0$.

(c) Find h_∞ and sketch a graph of $h(t)$.

(d) Discuss the model critically.

66. Logistic Growth Model. A simple model for the evolution of a population over time was $p' = a \cdot p$. Its solution showed an exponential behavior. Now let us consider an improved so-called logistic growth model, bearing in mind that the resources (such as habitat and food) are limited:

$$\frac{dp}{dt} = ap - bp^2.$$

This model was established by the Dutch mathematician Verhulst (1804–1849). The positive constants a, b are the so-called **vitality coefficients** of the population. Of course, $a \gg b > 0$, because for small populations the simple model suffices as an approximation. For larger p-values, the negative term $-bp^2$ plays an increasingly growth-inhibiting role.

When counting a population of bacteria in a test tube[12], one finds that the logistic model works excellently! Let the initial population be given by $p(0) = p_0$.

(a) By means of a direction field, sketch the solutions for all possible initial populations $p(0) = p_0$ and find the limit as $t \to \infty$ by inspecting the graphs.

(b) Verify that the general solution for $p_0 > 0$ is

$$p(t) = \frac{1}{\frac{b}{a} + De^{-at}},$$

and determine the D-values for each of the initial conditions

$$p_0 < \frac{a}{b}, \quad p_0 > \frac{a}{b}, \quad p_0 = \frac{a}{b}.$$

(c) Show that also

$$p(t) = p_0 \cdot \frac{a}{b \cdot p_0 + (a - b \cdot p_0) \cdot e^{-at}}$$

describes the entire set of solutions.

Remark: A more accurate study of the statistical data of the world population reveals that the logistic growth model is hardly useful. Nevertheless, diagrams taken over decades forecasting the population from the past to the future display an approximate S-shape.

---

[12] More on this in [5].

67. Chemical Reaction. Problem taken from [4]. Collisions of the following kind: A molecule of some substance P collides with a molecule of another substance Q, generating a molecule of a new substance Y.

Assuming that p and q are the associated initial concentrations of P and Q, and $y(t)$ is the concentration of Y at time t, then $p - y(t)$ and $q - y(t)$ are the correeponding concentrations of P and Q at time t.

The reaction rate is given by the equation

$$\frac{dy}{dt} = k \cdot [p - y(t)] \cdot [q - y(t)]$$

where k is a positive constant. Furthermore, $q > p > 0$ and $y(0) = 0$.

(a) Sketch the direction field and find $\lim_{t \to \infty} y(t)$ without explicitly computing $y(t)$.

(b) Compute $y(t)$.

68. Models for Tumor Growth. The following models are used for cancer therapies and serve as a reference in order to quantify their effectiveness.

(a) Experimental observations show that the volume V of free living dividing cells grows exponentially with time. Therefore

$$V'(t) = \alpha \cdot V(t) \quad \text{where} \quad \alpha > 0.$$

Compute the period of time T required for the volume to double.

(b) For hard tumors, however, the doubling time T increases with time, because the cells in the growing nucleus are no longer supplied with blood and eventually die. Investigations show that the measured values, up to 1000 times the original volume, are remarkably well described by the Gompertz[13] function

$$V(t) = V_0 \cdot e^{\frac{r}{s}(1 - e^{-st})}$$

with parameters $r, s > 0$. It has an S-shape similar to the logistic function. Determine the limit of $V(t)$ as $t \to \infty$ and, by differentiating, the differential equation of V.

Remark: The Gompertz function also plays a role in market or trend research.

69. Friction of Air. An object of given mass m moves along a horizontal line. At time $t = 0$ the initial velocity is v_0. Only a breaking force, caused by the friction of air, acts upon the point mass. It is proportional to the square of the velocity with the given proportionality factor $k > 0$.

Compute the following for the motion of the point mass (give answers in terms of the parameters m, v_0, k and simplify):

(a) the velocity function $v(t)$ varying with time,

(b) the position function $x(t)$ with $x(0) = 0$,

(c) the limits of velocity and position as $t \to \infty$.

---

[13] Benjamin Gompertz (1779–1865) was born into a Jewish family and a self-educated mathematician. He was a British citizen who became a Fellow of the Royal Society.

70. Orthogonal Trajectories. Compute the orthogonal trajectories of the family of curves $y = kx^3$ and describe them geometrically.

71. Smith Chart. Show analytically that the orthogonal trajectories of the family of circles K through the origin with centers on the y-axis are the family of circles K^* through the origin with centers on the x-axis, see figure.[14]

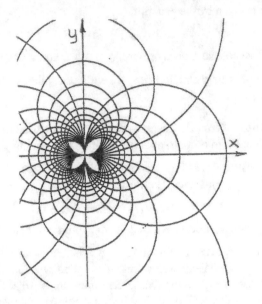

Step-by-step instructions:

(a) Solve the equation of the circles $x^2 + (y - c)^2 = c^2$ for c,

(b) differentiate the equation of the circles, solve for y', and substitute c,

(c) find a differential equation for the orthogonal family of curves,

(d) verify that K^* are the solutions by substituting into the differential equation.

---

[14] Application in electrical engineering.

Chapter 4
Second Order Differential Equations and Systems with Applications

Many differential equations in the natural sciences are of second order.
Here we generally do not care as much about solving techniques as about understanding them. Above all, we are interested in establishing differential equations for applications, that is, in practicing mathematical modeling.

4.1 Second Order Differential Equation

General form: $$y'' = f(t, y, y').$$

On the right, we have an "arbitrarily complicated" term in the variables t, y, y'.

Visualizing via a direction field is not possible!

The function $y(t)$ is a solution, if the equation $y''(t) = f(t, y(t), y'(t))$ is identically satisfied for all t-values.
Typically an IVP of second order has two initial conditions, because we need to integrate twice. This leads to two integration constants.

4.2 System of First Order Differential Equations

The general form
$$\left.\begin{array}{l} y_1' = f_1(x, y_1, y_2, \ldots, y_n) \\ y_2' = f_2(x, y_1, y_2, \ldots, y_n) \\ \cdots \qquad \cdots\cdots \\ y_n' = f_n(x, y_1, y_2, \ldots, y_n) \end{array}\right\}$$

with initial conditions
$$y_1(x_0) = a_1, \quad y_2(x_0) = a_2, \quad \ldots, \quad y_n(x_0) = a_n$$
describes an initial value problem involving n unknown functions $y_i(x)$.
The n functions $f_i = f_i(x, y_1, y_2, \ldots, y_n)$ and the n initial values a_i are given.

Hence, in general, the derivatives depend on various functions: The differential equations are coupled to one another.

Systems of first order differential equations are important also because **an n-th order differential equation in the unknown function y(x) may be represented equivalently by a system of n first order differential equations**.

To illustrate this, let us consider the general second order differential equation

$$y'' = f(t, y, y').$$

It may be rewritten as a system of first order differential equations in the two unknown functions $y(t)$ and $z(t) = y'(t)$:

$$\left. \begin{array}{rl} y' = & z \\ z' = & f(t, x, y) \end{array} \right\}.$$

Example 4.30 The second order differential equation

$$y'' = y^2 - y' \sin t + e^{-t}$$

may be reformulated as a system of first order differential equations:

$$\left. \begin{array}{rl} y' = & z \\ z' = & y^2 - z \cdot \sin t + e^{-t} \end{array} \right\}. \qquad \qquad \Diamond$$

4.3 Projectile Motion

This is a simple model that already involves a system of differential equations.

A particle is thrown obliquely into the air. If we neglect the effects of air resistance, then the particle begins its path with downward acceleration $\vec{g}$ due to its own weight $\vec{G}$, see figure.

Initial conditions at time $t = 0$:

- $x(0) = y(0) = 0$
- α = inclination angle at start,
  v_0 = initial speed

$$\vec{v_0} = \begin{pmatrix} v_{0x} \\ v_{0y} \end{pmatrix} = \begin{pmatrix} \cos \alpha \cdot v_0 \\ \sin \alpha \cdot v_0 \end{pmatrix}.$$

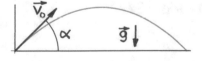

As usual, $\vec{r}$ is the position, $\vec{v}$ the velocity, and $\vec{a}$ the acceleration of the motion.

Then

$$\vec{r}(t) = \begin{pmatrix} x(t) \\ y(t) \end{pmatrix}, \qquad \vec{v}(t) = \begin{pmatrix} \dot{x}(t) \\ \dot{y}(t) \end{pmatrix}, \qquad \vec{a}(t) = \begin{pmatrix} \ddot{x}(t) \\ \ddot{y}(t) \end{pmatrix}.$$

Newton's equation of motion

$$m \cdot \ddot{\vec{r}}(t) = \vec{G} = m \cdot \vec{g}$$

in Cartesian coordinates, after reducing m implies

$$\begin{pmatrix} \ddot{x}(t) \\ \ddot{y}(t) \end{pmatrix} = \begin{pmatrix} 0 \\ -g \end{pmatrix} \quad \text{where } g = 9.81 \text{m/s}^2 = \text{ acceleration of gravity.}$$

Here we have an uncoupled system of differential equations that can be solved by simply integrating twice with regard to t.
Imposing the initial conditions yields

$$\vec{v}(t) = \begin{pmatrix} \dot{x}(t) \\ \dot{y}(t) \end{pmatrix} = \begin{pmatrix} v_{0x} \\ -gt + v_{0y} \end{pmatrix}.$$

The horizontal component of the velocity is constant, the vertical component changes linearly with t.
After another integration and assuming that the initial position is at the origin, we obtain

$$\vec{r}(t) = \begin{pmatrix} x(t) \\ y(t) \end{pmatrix} = \begin{pmatrix} v_{0x} \cdot t \\ v_{0y} \cdot t - \frac{g}{2} \cdot t^2 \end{pmatrix} = \begin{pmatrix} \cos\alpha \cdot v_0 \cdot t \\ \sin\alpha \cdot v_0 \cdot t - \frac{g}{2} \cdot t^2 \end{pmatrix}.$$

Eliminating t by replacing $t = x/v_{0x}$ from the x-component into the y-component yields a quadratic relation

$$y = \frac{v_{0y}}{v_{0x}} \cdot x - \frac{g}{2v_{0x}^2} \cdot x^2,$$

which describes a parabola.

The time t_W until the projectile reaches the ground is obtained as follows:

$$y(t) = t \cdot \left(v_{0y} - \frac{g}{2} t \right) = 0 \quad \Longrightarrow \quad t_W = \frac{2v_{0y}}{g}.$$

The horizontal range of the particle is

$$x_W = x(t_W) = \frac{2}{g} v_{0x} v_{0y} = \frac{v_0^2}{g} 2\cos\alpha \sin\alpha = \frac{v_0^2}{g} \sin(2\alpha).$$

Thus the maximum horizontal range $x_{max} = \dfrac{v_0^2}{g}$ is obtained for an initial inclination angle of $45°$.

The height of the culmination point (i.e. the vertex of the parabola) is, therefore,

$$y_{max} = y\left(\frac{t_W}{2} \right) = \frac{v_{0y}^2}{2g}.$$

Remark: Range, height, and time of the trajectory on the moon are about six times those on the earth, because

$$g_{moon} \approx \frac{g_{earth}}{6}.$$

4.4 Modeling With Air Drag

The air drag $\vec{W}$ obeys the law

$$|\vec{W}| = k \cdot |\vec{v}|^\lambda \text{ where } 1 < \lambda \le 2.$$

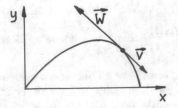

With the same notations $\vec{r}, \vec{v}, \vec{a}$ as in the
previous example, we obtain

$$m \cdot \vec{a} = \vec{W} + \vec{G},$$

where the resultant force is on the right-hand side. The air drag $\vec{W}$ is directed opposite the velocity $\vec{v}$. Thus

$$\vec{W} = -k \cdot |\vec{v}|^{\lambda-1} \cdot \vec{v} = -k(\dot{x}^2 + \dot{y}^2)^{\frac{1}{2}(\lambda-1)} \cdot \vec{v}.$$

Therefore, dividing by m yields the equations of motion in terms of coordinates

$$\left. \begin{array}{l} \ddot{x} = \quad -\frac{k}{m}(\dot{x}^2 + \dot{y}^2)^{\frac{1}{2}(\lambda-1)} \cdot \dot{x} \\ \ddot{y} = -g - \frac{k}{m}(\dot{x}^2 + \dot{y}^2)^{\frac{1}{2}(\lambda-1)} \cdot \dot{y} \end{array} \right\}. \tag{4.1}$$

Together with the four initial conditions for position and velocity

$$x(0) = x_0 \quad y(0) = y_0 \quad \dot{x}(0) = v_{0x} \quad \dot{y}(0) = v_{0y},$$

this gives rise to an initial value problem of two coupled nonlinear second order
differential equations. For the special case $\lambda = 2$, the exponent $= \frac{1}{2}$.

4.5 Coriolis Force in Meteorology

Global circulation of the winds on the rotating earth is an effect of the so-called
Coriolis force. It governs the time-varying velocity $\vec{w}(t)$ of a parcel of air and consequently also its path $\vec{r}(t)$[1]. In the computations that follow we assume the ideal
case that the considered region in the northern hemisphere has constant pressure and
that friction is negligible (which is not true near the ground).
Let the air parcel at point P on the earth's surface have an initial speed v_0 directed
to the north.
We start from a Cartesian coordinate system with origin at P and the coordinate axes
pointing to the east (x-direction) and to the north (y-direction). Furthermore, let $u(t)$
be the coordinate of $\vec{w}$ in eastern direction, $v(t)$ the one in northern direction:

$$\vec{w}(t) = \begin{pmatrix} u(t) \\ v(t) \end{pmatrix} \text{ with initial condition } \vec{w}(0) = \vec{w_0} = \begin{pmatrix} 0 \\ v_0 \end{pmatrix}. \tag{4.2}$$

---

[1] Dr. med. dent. Manfred Jenni inspired this contribution. He did not hesitate to delve with me into
the mathematical depths of partial differential equations and atmospherical physics.

Due to the Coriolis force for the geographical latitude Φ, the equation of motion implies the following system of differential equations (accelerations on the left):

$$\left.\begin{array}{l} \dot{u}(t) = f \cdot v(t) \\ \dot{v}(t) = -f \cdot u(t) \end{array}\right\} \qquad \text{where } f = 2\Omega \sin(\Phi). \tag{4.3}$$

Here $\Omega = 2\pi/24\text{h} = 7.272 \cdot 10^{-5}/\text{s}$ is the angular velocity of the earth's rotation.

In the northern hemisphere Φ and f are both positive, in the southern hemisphere they are negative. The south pole has latitude $\Phi = -90°$.

If we analyze just a small region in the northern hemisphere, then the geographical latitude is about constant. Together with the initial conditions, we obtain the solutions of (4.3)

$$\vec{w}(t) = \begin{pmatrix} u(t) \\ v(t) \end{pmatrix} = v_0 \begin{pmatrix} \sin(f \cdot t) \\ \cos(f \cdot t) \end{pmatrix} = \begin{pmatrix} \dot{x} \\ \dot{y} \end{pmatrix},$$

as is easily verified. Evidently, this velocity vector always has the same magnitude v_0 (due to energy conservation).

By integrating the above equation and imposing the initial conditions $x(0) = 0$, $y(0) = 0$ we obtain the equation of a circular motion of the air parcel:[2]

$$\vec{r}(t) = \begin{pmatrix} x(t) \\ y(t) \end{pmatrix} = \frac{v_0}{f} \begin{pmatrix} 1 - \cos(f \cdot t) \\ \sin(f \cdot t) \end{pmatrix}$$

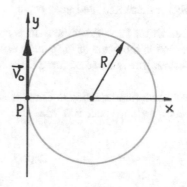

In the northern hemisphere ($f > 0$), the air parcel starts at the origin P and turns along the circle with radius $R = \dfrac{v_0}{f}$ in clockwise direction.

In the southern hemisphere ($f < 0$), however, it revolves in counterclockwise direction, see figure.

Notice:

Owing to Equation (4.3), the angular frequency f and hence also the period $T = \frac{2\pi}{f}$ depend only on Φ, not on R.

More generally, if the pressure is not constant, then the following holds: **In the northern hemisphere**, the Coriolis force causes a moving object to be **deflected to the right** relative to its direction of travel. **In the southern hemisphere**, it causes a **deflection to the left.**

In a high pressure area of the northern hemisphere, a deflection to the right causes a clockwise rotation, in a low pressure area it causes a counterclockwise rotation. In the southern hemisphere the converse is true. More on this in [25].

The more north or south the point P, the smaller the radius R and the period T for a fixed v_0. Near the equator there is practically no Coriolis force.

---

[2] In meteorology it is customary to denote by f the angular frequency. In physics, however, as a rule f denotes the frequency and ω the angular frequency.

Example 4.31
At a latitude of $\Phi \approx 45°$, the angular frequency is $f = 1.03 \cdot 10^{-4}$ s^{-1}. In this case a simple relation holds:

$$R \approx 10^4 \cdot v_0 \quad \text{where } [R] = m, \quad [v_0] = \frac{m}{s},$$

with period $T = \dfrac{2\pi}{f} \approx 62800$ s ≈ 17.5 h, irrespective of R.

Concrete cases:

- For a wind velocity of $v_0 = 10$ m/s, the radius is $R = 100$ km.
- Interestingly, the computations are valid also for an unperturbed ocean current. For a typical flow velocity of $v_0 = 0.1$ m/s, the radius is $R = 1$ km. $\diamond$

4.6 Vector Fields and Field Lines

Vector fields and field lines occur in various fields of physics.

Definition 12. *A **vector field** is characterized by the property that each point of a region in the plane or in space is associated with a vector that is attached to it.*
*A **field line** is a line or curve that at each point is tangential to the vector field.*

The analytical equation in Cartesian coordinates of a stationary (i.e. time-independent) vector field in space is of the form

$$\vec{v}(x,y,z) = \begin{pmatrix} v_1(x,y,z) \\ v_2(x,y,z) \\ v_3(x,y,z) \end{pmatrix}.$$

Here x, y, z are the coordinates of point $P(x,y,z)$ at which the vector is attached, and v_1, v_2, v_3 are scalar coordinate functions in the three variables x, y, z.
Of course, in a plane vector field z and v_3 are omitted.

In Section 1.4.3 about tangents of curves we found that the vector derivative of a field line $\vec{r}(s)$ with respect to some real parameter s is always tangent to the field line.
With this in mind, the system of first order differential equations for the field lines is

$$\vec{r}\,'(s) = \begin{pmatrix} x' \\ y' \\ z' \end{pmatrix} = \begin{pmatrix} v_1(x,y,z) \\ v_2(x,y,z) \\ v_3(x,y,z) \end{pmatrix},$$

where $x = x(s)$, $y = y(s)$, $z = z(s)$ and $'$ denotes the derivative with respect to s.

Remark: We deliberately did not use t as a parameter, because s may not represent the time.

For the special field line that passes through the given point

$$P(x(s_0), y(s_0), z(s_0)) = P(x_0, y_0, z_0),$$

there are three initial conditions and hence we have an Initial value problem.

For a plane vector field

$$\vec{v}(x,y) = \begin{pmatrix} v_1(x,y) \\ v_2(x,y) \end{pmatrix},$$

let us introduce the function $y(x)$ such that the derivative corresponds to the slope of the vector $\vec{v}(x,y)$ at point (x,y). Thus we obtain the differential equation

$$\frac{dy}{dx} = \frac{v_2(x,y)}{v_1(x,y)} = f(x,y), \tag{4.4}$$

thus gaining a different approach to computing the field lines $y(x)$.

We may visualize (4.4) via a direction field by computing the slopes $f(x,y)$ for a dot screen and then drawing the corresponding "compass needles".

Computer algebra systems visualize vector fields in space or in a plane by displaying the "compass needles" as well as the field lines. Examples thereof are the diagram of the differential equation (2.1), or the diagram in the predator-prey problem discussed later.

Direction fields are usually 2-dimensional, and vector fields are used in two- or three-dimensional regions.

4.7 Helmholtz Coils

The figure shows a circular loop of radius $r = 1$ carrying a current i. We wish to compute the induced magnetic vector field $\vec{H}$ at some arbitrary point $P(x,y,z)$.

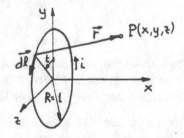

Biot-Savart's Law states

$$\vec{H} = \frac{i}{4\pi} \oint_C \frac{\vec{d\ell} \times \vec{r}}{r^3},$$

where $\vec{d\ell}$ is the current element and r the distance from the current element to point P.

As the vector field has rotational symmetry about the x-axis, it suffices to examine the vector field in the (x,y)-plane (where $z = 0$).
Parametrizing the conductor in the (y,z)-plane by way of the parameter t yields

$$\vec{\ell}(t) = \begin{pmatrix} 0 \\ \cos t \\ \sin t \end{pmatrix} \quad \text{with current element} \quad \vec{d\ell} = \dot{\vec{\ell}} \cdot dt = \begin{pmatrix} 0 \\ -\sin t \\ \cos t \end{pmatrix} \cdot dt.$$

The distance vector in the (x,y)-plane is

$$\vec{r}(t) = \begin{pmatrix} x \\ y \\ 0 \end{pmatrix} - \begin{pmatrix} 0 \\ \cos t \\ \sin t \end{pmatrix} = \begin{pmatrix} x \\ y - \cos t \\ -\sin t \end{pmatrix},$$

where $r^2 = x^2 + (y - \cos t)^2 + \sin^2 t = x^2 + y^2 - 2y\cos t + 1$.

$$\vec{d\ell} \times \vec{r} = \begin{pmatrix} 0 \\ -\sin t \\ \cos t \end{pmatrix} \times \begin{pmatrix} x \\ y - \cos t \\ -\sin t \end{pmatrix} \cdot dt = \begin{pmatrix} 1 - y\cos t \\ x\cos t \\ x\sin t \end{pmatrix} \cdot dt.$$

Accordingly,

$$\vec{H}(x,y,0) = \begin{pmatrix} H_x \\ H_y \\ H_z \end{pmatrix} = \frac{i}{4\pi} \oint_0^{2\pi} \frac{1}{r^3} \cdot \begin{pmatrix} 1 - y\cos t \\ x\cos t \\ x\sin t \end{pmatrix} \cdot dt.$$

The components of the magnetic field $\vec{H}$ are

$$H_x = \frac{i}{4\pi} \int_0^{2\pi} \frac{1 - y\cos t}{(x^2 + y^2 - 2y\cos t + 1)^{3/2}} \cdot dt,$$

$$H_y = \frac{i}{4\pi} \int_0^{2\pi} \frac{x\cos t}{(x^2 + y^2 - 2y\cos t + 1)^{3/2}} \cdot dt,$$

$$H_z = 0.$$

The last component vanishes due to rotational symmetry.
These integrals can no longer be expressed analytically in closed form. Hence it is advisable to numerically compute the two integrals H_x and H_y at each point (x,y).

The special case of an $\vec{H}$-field on the axis of rotation, i.e. where $y = 0$, for a point at distance x of the circular loop is

$$H_x = \frac{i}{4\pi} \frac{1}{(1+x^2)^{3/2}} \int_0^{2\pi} 1 \cdot dt = \frac{i}{2} \frac{1}{(1+x^2)^{3/2}}$$

with the maximum value $i/2$ at the center of the circle.

If there is a second circular loop parallel to the first one shifted along the axis of symmetry at $x = a$, then the following results:

In the integrals for H_x and H_y, the variable x must be replaced by $(a + x)$.

If now two or three circular coils are arranged co-axially, then they are called Helmholtz coils.[3] The resultant $\vec{H}$-field of such an arrangement is the superposition of the individual fields, i.e. their vector sum.

The students Schafroth and Giezendanner[4] computed and represented graphically the field lines of 1, 2, and 3 Helmholtz coils:

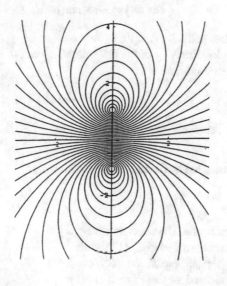

---

[3] Hermann von Helmholtz (1821-1894), German professor of physics and physiology, was a scholar of universal knowledge.

[4] Project conducted in a mathematics workshop of the Department of Technology and Computer Science at the Bern University of Applied Sciences.

4.8 Oscillations and Resonance

Example 4.32 Let us determine the general solutions of the linear homogeneous differential equation

$$y'' = -\omega^2 y.$$

The solutions are

$$y_1(t) = \sin(\omega t) \qquad y_2(t) = \cos(\omega t).$$

The reason is that

$$\frac{d^2}{dt^2} \sin(\omega t) = -\omega^2 \sin(\omega t).$$

An analogous equation holds for $\cos(\omega t)$.

Due to homogeneity and linearity of the differential equation, we obtain the general solution as a linear combination

$$y(t) = a\cos(\omega t) + b\sin(\omega t) \tag{4.5}$$

with two integration constants a and b. $\diamond$

Example 4.33 Harmonic Oscillator.

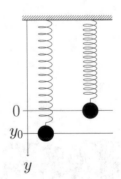

Let us analyze the motion of the mass m attached to a spring with linear characteristics, see figure. Let D be the spring constant measured in N/cm. The point mass is released at time $t = 0$ from a displacement y_0 relative to the equilibrium position $y = 0$.

Newton's law of motion states

$$m\ddot{y} = -D \cdot y.$$

With the initial conditions $y(0) = y_0$ and $\dot{y}(0) = 0$, the solution becomes

$$y(t) = y_0 \cdot \cos(\omega t) \qquad \text{with angular frequency} \quad \omega = \sqrt{\frac{D}{m}}.$$

The harder the spring, the larger the frequency.
The larger the mass, the smaller the frequency.

Remark: Weight and preloaded spring force in the equilibrium position cancel each other, whence were omitted. $\diamond$

Example 4.34 Pendulum

A mass m is attached to a massless string or rod of length ℓ practically without friction. If the mass is displaced from its equilibrium position, then it performs an oscillatory motion. We shall study it by establishing and solving a differential equation.

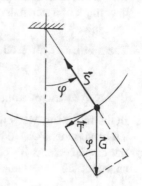

Our goal is to compute the angle $\varphi(t)$ as a function of time.

The resultant restoring force $\vec{T}$ acts tangentially to the circular arc, because the string takes the weight component $\vec{S}$ in its direction.

The arclength is $\ell \cdot \varphi(t)$, hence the velocity $v(t) = \ell \cdot \dot{\varphi}(t)$, and the acceleration $\dot{v}(t) = a(t) = \ell \cdot \ddot{\varphi}(t)$.

Newton's law of motion states

$$m \cdot \ell \cdot \ddot{\varphi}(t) = -\sin(\varphi) \cdot m \cdot g.$$

The minus sign accounts for the fact that if the angle is positive, then the force T acts to the left; if the angle is negative, T acts to the right. That is, angle and resultant force are always opposite to each other and therefore have different signs. By dividing both sides by $m \cdot \ell$ we get

$$\ddot{\varphi}(t) = -\frac{g}{\ell} \sin(\varphi). \tag{4.6}$$

This is a nonlinear differential equation that cannot be solved by way of a closed expression! Obviously, the mass has no influence on the motion at all.

- If the **displacement is small** (e.g. $\varphi < 30°$), then $\sin \varphi \approx \varphi$. Hence the differential equation becomes linear: $\ddot{\varphi}(t) = -\frac{g}{\ell} \varphi$.
  Introducing the constant $\omega^2 = \frac{g}{\ell}$ gives
  $$\ddot{\varphi} = -\omega^2 \cdot \varphi.$$

According to (4.5), the general solution is $\varphi(t) = a\cos(\omega t) + b\sin(\omega t)$.
The coefficients a and b are determined by the initial position $\varphi(0) = \varphi_0 = a$ and the initial velocity $v(0) = v_0 = \ell \cdot \dot{\varphi}(0) = \ell b \omega \Rightarrow b = \dfrac{v_0}{\ell \omega}$. This produces the solution

$$\varphi(t) = \varphi_0 \cos(\omega t) + \frac{v_0}{\ell \omega} \sin(\omega t) \quad \text{with angular frequency} \quad \omega = \sqrt{\frac{g}{\ell}}.$$

The smaller the length ℓ, the larger the angular frequency ω. Equation (3.10) implies

$$\varphi(t) = A \cdot \sin(\omega t + \alpha) \quad \text{with amplitude} \quad A = \sqrt{\varphi_0^2 + \left(\frac{v_0}{\ell \omega}\right)^2}.$$

Remark: In the book [13], Section B.3.2 analyzes a pendulum with mass m attached to a spring (elastic pendulum).

- Now let us look at the problem with **large displacements**. Instead of a string, we may think of a thin massless metal rod. Then initial displacements up to $180°$ are possible.

  The **nonlinear differential equation of second order**

  $$\ddot{\varphi} = -\frac{g}{\ell}\sin\varphi$$

  is solved by use of a numerical method.

  In the following Mathematica code[5], $\ell = 1$ and φ is renamed w.
  The assignment was to generate a diagram consisting of solutions for the 19 initial displacements $0°, 10°, 20°, 30°, \ldots, 180°$. Here the initial displacements are spaced $\pi/20$, i.e. $9°$ apart.

  (a) Define the constants:
  ```
  h=20; g=9.81; (*g=acceleration of gravity*)
  ```

  (b) Solve the differential equation numerically:
  ```
  schwing[ampl_] := NDSolve[w″[t] + g * Sin[w[t]] == 0,
  w[0] = ampl, w′[0] == 0, w, {t, 0, 4}];
  ```
  The quantity ampl_ is the parameter.

  (c) Generate a list for various initial displacements:
  ```
  liste = Flatten[Table[w[t] /. schwing[x], {x, 0, Pi, Pi/h}]];
  ```

  (d) Generate the diagram below:
  ```
  arrayplot = Map[Plot[#, {t, 0, 2}], PlotStyle- > {Thickness[0.00001]},
  DisplayFunction- > Identity]&, liste];
  Show[{arrayplot}, DisplayFunction- > $DisplayFunction];
  ```

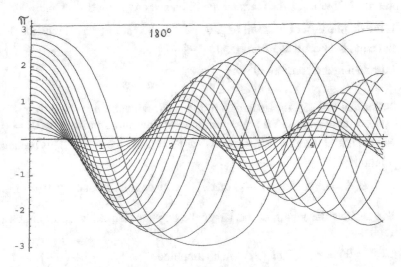

---
[5] Project carried out by my former students Schafroth and Giezendanner in a mathematics workshop of the Department of Technology and Computer Science at the Bern University of Applied Sciences.

Interpretation: It is readily seen that for the first curves with amplitudes up to about 30° the periods are practically constant. The periods increase with the amplitudes, thus distorting the shape of the sine curve. The extreme cases with initial conditions 0 and π yield constant solutions: $\varphi = 0$ is stable, $\varphi = \pi$ is unstable ◊

Example 4.35 Electromagnetic LC Oscillation Circuit.

Interestingly, we shall see that the electromagnetic LC oscillation circuit behaves analogously to the mechanical harmonic oscillator. Let the capacitor carry an initial charge q_0. The circuit is to be closed at time $t = 0$.

Kirchhoff's law states that the sum of the voltages

$$u_C(t) = q(t)/C \text{ and } u_L(t) = L \cdot \frac{di}{dt}$$

disappears:

$$\frac{1}{C} \cdot q + L \cdot \frac{di}{dt} = 0.$$

The rate of change of the charge is the current: $\dot{q}(t) = i(t)$.

Differentiating the equation yields

$$\frac{1}{C} i + L \frac{d^2 i}{dt^2} = 0 \quad \Longleftrightarrow \quad \frac{d^2 i}{dt^2} = -\frac{1}{LC} \cdot i.$$

Initial conditions: $u(0) = \dfrac{q_0}{C} = u_0$ and $i(0) = 0$,

The solution of the IVP is

$$i(t) = k \cdot \sin(\omega t)$$

with angular frequency $\omega = 1/\sqrt{LC}$.
Now let us determine the constant k. By equating the coefficients in

$$u(t) = u_L(t) = L \cdot \frac{di}{dt} = L\omega k \cos(\omega t) = u_0 \cos(\omega t)$$

we obtain $k = \dfrac{u_0}{\omega L}$. Therefore,

$$u(t) = u_0 \cos(\omega t) \quad i(t) = \frac{u_0}{\omega L} \sin(\omega t).$$

The phase angle between voltage and current is 90°.   ◊

Example 4.36 Weakly Damped Free Oscillator:

In the real world, moving pendulums are subject to weak air resistance that we have neglected so far. In a system with brake fluid,
see figure, the influence of friction is no
longer negligible.

Let us assume that the force of resistance is
proportional to the velocity $\dot{y}$.

Then Newton's law has an additional damping term:

$$m\ddot{y} = -D \cdot y - c \cdot \dot{y}.$$

Dividing by m gives rise to

$$\ddot{y} + 2\rho \cdot \dot{y} + \omega_0^2 \cdot y = 0 \quad \text{where } \omega_0^2 = D/m, \quad 2\rho = \frac{c}{m}.$$

Remark: The factor 2 simplifies computations.

Initial position and initial velocity are $y(0) = y_0, \quad \dot{y}(0) = v_0.$

Without proof we give the solution for an oscillator in the case of weak damping,
i.e. $\omega_0^2 > \rho^2$:

$$y(t) = e^{-\rho t} \cdot [y_0 \cos(\omega t) + \frac{v_0 + \rho y_0}{\omega} \sin(\omega t)] \qquad \text{where } \omega^2 = \omega_0^2 - \rho^2. \quad (4.7)$$

It can be rewritten as
$$y(t) = Ae^{-\rho t} \cdot \sin(\omega t + \alpha) \quad \text{where } A = \sqrt{y_0^2 + \left(\frac{v_0 + \rho y_0}{\omega}\right)^2}.$$

This is an exponentially damped oscillation:

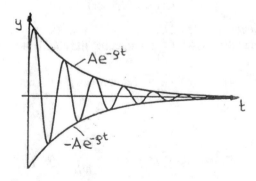

The quantity ρ is decisive for the speed of decay. The frequency ω of the damped
oscillator is slightly smaller than ω_0. For small ρ, we have $\omega \approx \omega_0$.

Checking the solution: For $\rho = 0$, the answer is that of an undamped oscillator. $\Diamond$

Example 4.37 Forced Oscillation, Resonance Phenomenon

In addition to the weak damping we have here an **external periodic driving force** acting on the system of oscillators. For example, take the driving force $K(t) = K_0 \cos(\Omega t)$ with amplitude K_0 and driving angular frequency Ω (see figure). The associated nonhomogeneous linear differential equation is

$$\ddot{y} + 2\rho \cdot \dot{y} + \omega_0^2 \cdot y = \frac{K_0}{m} \cos(\Omega t). \qquad (4.8)$$

Let us restrict ourselves to the **steady-state behavior of the system**: That is, the behavior after a short initial oscillatory motion, which is to be disregarded. During the initial transient process there are additional terms that approach zero exponentially, similar to the free damped oscillator in the previous example.

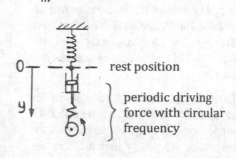

0 — rest position

y

periodic driving force with circular frequency

As will be seen, the initial conditions have no influence on the steady-state component y_{st} of the solution, which is given here without proof [6]:

$$y_{st} = k \cdot \left\{ \frac{\omega_0^2 - \Omega^2}{N} \cos(\Omega t) + \frac{2\rho\Omega}{N} \sin(\Omega t) \right\} = A(\Omega) \cdot \cos[\Omega t + \varphi(\Omega)] \qquad (4.9)$$

$$\text{where} \quad N = (\omega_0^2 - \Omega^2)^2 + 4\rho^2\Omega^2 \text{ and } k = \frac{K_0}{m}.$$

Our aim is to analyze the amplitude $A(\Omega)$ as a function of the driving angular frequency Ω.

The sum of the squares of the two numerators is N. Therefore,

$$A(\Omega) = \frac{k}{\sqrt{N}} = \frac{k}{\sqrt{(\omega_0^2 - \Omega^2)^2 + 4\rho^2\Omega^2}}.$$

Checking the dimensions shows that they are consistent:

$$[\rho] = \frac{[c]}{[m]} = \frac{N}{(m/s) \cdot kg} = 1/s.$$

The amplitude is a maximum, if the denominator and hence its square is a minimum:

$$\frac{d}{d\Omega}[(\omega_0^2 - \Omega^2)^2 + 4\rho^2\Omega^2] = 2(\omega_0^2 - \Omega^2) \cdot (-2\Omega) + 8\rho^2\Omega = 0.$$

Dividing by 4 and factoring out Ω gives

$$\Omega(2\rho - \omega_0^2 + \Omega^2) = 0 \iff \Omega_1 = 0 \text{ or } \Omega^2 = \omega_0^2 - 2\rho^2.$$

---

[6] The initial transient process can be computed by aid of the Laplace transform.

Hence the resonance frequency Ω_{res} and the resonance amplitude A_{res} are

$$\Omega_{\text{res}} = \sqrt{\omega_0^2 - 2\rho^2}, \qquad A_{\text{res}} = A(\Omega_{\text{res}}) = \frac{k}{2\rho\sqrt{w_0^2 - \rho^2}}.$$

In the case of weak damping ($\rho \ll \omega_0$), the resonance frequency is close to the frequency ω_0 of the free undamped oscillator. Furthermore, it is evident from the equation that A_{res} may peak extremely. This is called a **resonance disaster**:

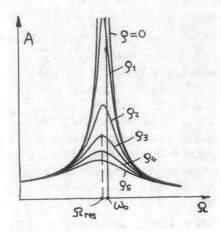

Even though the amplitude K_0 of the external driving force may be small, the amplitude $A(\Omega)$ in the vicinity of the resonance frequency may build up until the system collapses.

The figure shows amplitude curves for various ρ.
In the undamped extreme case $\rho = 0$, the curve has a singularity at ω_0. Here the amplitude grows to infinity.

Let us now investigate the **phase angle** $\varphi(\Omega)$ by considering the ratio of the coefficients of cosine and sine in (4.9). It suffices to look at the phase angle of the following auxiliary expression:

$$h(t) = (\omega_0^2 - \Omega^2)\cos(\Omega t) + (2\rho\Omega)\sin(\Omega t).$$

We begin by looking at the following three cases:

(a) Ω very small. Then $h(t) \approx \omega_0^2 \cos(\Omega t)$. Hence h, and thus, y_{st} is practically in phase with the driving force $K(t) = K_0 \cos(\Omega t)$. Therefore, $\varphi(\Omega) \approx 0$.

(b) $\Omega \to \infty$. Then the quotient $\dfrac{(\omega_0^2 - \Omega^2)}{(2\rho\Omega)}$ of the two coefficients of the function $h(t)$ tends to $-\infty$. Hence h and y_{st} are practically in counterphase with the driving force $K(t) = K_0 \cos(\Omega t)$. Therefore, $\varphi(\infty) = \pi$.

(c) $\Omega = \omega_0$. Then $h(t) = 2\rho\omega_0 \cdot \sin(\omega_0 t)$ and we have a phase angle of $\varphi(\omega_0) = \frac{\pi}{2}$ relative to the driving force $K(t) = K_0 \cos(\Omega t)$.

In addition, let us look at the undamped case $\rho = 0$:

$$y_{\text{st}}(t) = \frac{k}{\omega_0^2 - \Omega^2} \cdot \cos(\Omega t).$$

Here the fraction changes its sign suddenly when Ω passes through ω_0: The function $\varphi(\Omega)$ changes discontinuously from 0 to π.

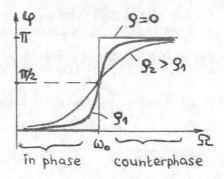

The diagram shows a so-called **phase jump**: For small damping, the curve asymptotically approaches the rectangular function where $\rho = 0$: Within a small range the phase angle jumps almost π units.

This model is the simplest case of a resonance phenomenon. More complicated oscillatory systems in practice have several resonance frequencies. It is advised to avoid or to drive through them quickly.   ◇

Examples of resonance phenomena:

(a) Tacoma Narrows Bridge in the U.S.A.: The wind-induced forces set the suspension cables oscillating with almost resonance frequency. The bridge collapsed. More on this in [2].
(b) Dance stages can collapse.
(c) When driving at a certain speed, the windows of the car may rattle.
(d) Using a small magnet attached to a rope, a heavy metal ball of several tons may be displaced arbitrarily, if one pulls the rope once each cycle.
(e) In large rotational engines, such as gas or water turbines or diesel motors in ships, critical frequencies must be avoided.

Example 4.38 Playful Experiment with Resonance.

The figure shows an oscillator consisting of a spring and a mass that is periodically incited by hand at the upper end of the spring. As the friction of air is minor, the phase jump is sensed the moment the incitation frequency Ω transits the natural frequency ω_0.

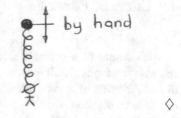

◇

Example 4.39 Electric Oscillations, Resonance.

As indicated earlier, there is a complete analogy between the forced mechanical oscillator with weak damping and the electric RCL circuit (see figure) with resistance R of the resistor, inductance L of the coil and capacitance C of the capacitor.

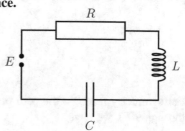

Let $E(t) = U\cos(\Omega t)$ be the voltage of an external power supply, where U is the amplitude and Ω is the angular frequency.

For the three voltages $U_R = R \cdot i$, $U_L = L \cdot \frac{di}{dt}$, $U_C = \frac{q}{C}$ with charge $q(t)$ and current $i(t)$, Kirchhoff's law $U_L + U_R + U_C = E(t)$ holds.

Substituting $i = \dfrac{dq}{dt}$ yields an equation for the charge:

$$L\ddot{q} + R\dot{q} + \frac{1}{C}q = E.$$

Dividing by L gives

$$\ddot{q} + \frac{R}{L}\ddot{q} + \frac{1}{LC}q = \frac{U}{L}\cos(\Omega t).$$

Mathematically this differential equation corresponds to Equation (4.8) of the mechanical problem.

Remark: All results of the mechanical oscillator, resonance included, apply, if in the results one consistently substitutes

$$\rho = \frac{R}{2L}, \quad \omega_0 = \frac{1}{\sqrt{LC}}, \quad k = \frac{U}{L}.$$

Weak damping exists, if and only if $\dfrac{R}{2L} < \dfrac{1}{\sqrt{LC}}$. $\Diamond$

4.9 Curve of Pursuit

At each corner of a square a dog starts running, see figure. Each dog runs in the instant direction of the one ahead.
Let us examine their paths.
We look at dog 1 chasing dog 2. Their positions are described by

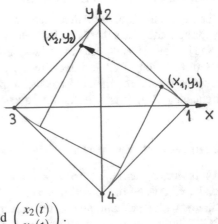

$$\begin{pmatrix} x_1(t) \\ y_1(t) \end{pmatrix} \text{ and } \begin{pmatrix} x_2(t) \\ y_2(t) \end{pmatrix}.$$

For symmetry reasons (the path of dog 2 is obtained by rotating the path of dog 1 through 90°), we have

$$\begin{pmatrix} x_2(t) \\ y_2(t) \end{pmatrix} = \begin{pmatrix} -y_1(t) \\ x_1(t) \end{pmatrix}.$$

The velocity of dog 1 has the direction of the difference of the two positions:

$$\begin{pmatrix} x'_1 \\ y'_1 \end{pmatrix} = \begin{pmatrix} x_2 \\ y_2 \end{pmatrix} - \begin{pmatrix} x_1 \\ y_1 \end{pmatrix} = \begin{pmatrix} x_2 - x_1 \\ y_2 - y_1 \end{pmatrix} = \begin{pmatrix} -x_1 - y_1 \\ x_1 - y_1 \end{pmatrix}.$$

For convenience, we adopt the notation $x_1 = x$ and $y_1 = y$. Then the path of dog 1 is represented by the following system of linear homogeneous differential equations:

$$\left. \begin{matrix} x' = -x - y \\ y' = x - y \end{matrix} \right\}.$$

With the initial conditions $(x(0), y(0)) = (1, 0)$ we obtain the solution

$$\vec{r}(t) = \begin{pmatrix} x(t) \\ y(t) \end{pmatrix} = e^{-t} \begin{pmatrix} \cos t \\ \sin t \end{pmatrix},$$

as can be verified by substitution.

The path is a **logarithmic spiral** $r(\varphi) = e^{-\varphi}$ with parameter $a = -1$.

It is known (see Exercise 28) that the constant angle φ between the position vector and its derivative in direction of the tangent is given by

$$\cos(\varphi) = \frac{a}{\sqrt{1 + a^2}} = \frac{-1}{\sqrt{2}} \implies \varphi = 135°.$$

Let us check this result at the initial point $(1, 0)$: The angle between the position vector of dog 1 and its instantaneous velocity in direction of dog 2 is indeed $180° - 45° = 135°$.

In the following we compute the length of the spiral.
Because

$$dx = \frac{dx}{dt} \cdot dt, \qquad dy = \frac{dy}{dt} \cdot dt \implies d\ell = \sqrt{dx^2 + dy^2} = \sqrt{\dot{x}^2 + \dot{y}^2}\, dt$$

and

$$c = \cos t, \quad s = \sin t, \quad \frac{dx}{dt} = e^{-t}(-c - s), \quad \frac{dy}{dt} = e^{-t}(c - s),$$

we obtain

$$d\ell = e^{-t} \sqrt{(-c - s)^2 + (c - s)^2}\, dt = e^{-t} \sqrt{2(c^2 + s^2)}\, dt = e^{-t} \sqrt{2}\, dt$$

and finally the length

$$\ell = \sqrt{2} \cdot \int_0^\infty e^{-t} \cdot dt = \sqrt{2}.$$

Even though the length is finite, the logarithmic spiral winds infinitely many times around its center (without ever reaching it)!

The following drawing of finitely many steps was done by Lok Fey Chan[7], then aged 12, during a course for gifted children in the Canton of Berne:

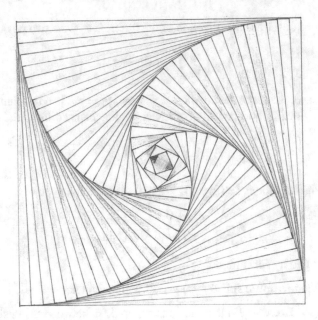

4.10 Coupled Pendulums

Let us consider two identical pendulums with the same lengths ℓ and the same masses m that are coupled by a spring. We recall that for an ordinary pendulum, the time-varying angle $\varphi(t)$ satisfies

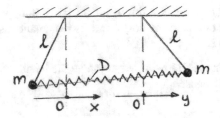

$$\ddot{\varphi} = -\frac{g}{\ell}\varphi,$$

if displacements are small.

Let us consider the horizontal displacement $x(t)$: For small displacements we may practically equate the arclength with $x(t)$: $x(t) = \ell \cdot \varphi(t)$.

Hence

$$\ddot{x}(t) = \ell \cdot \ddot{\varphi}(t) = -g\varphi(t) = -\frac{g}{\ell} \cdot x(t).$$

Or simply

$$\ddot{x} = -\omega_0^2 \cdot x, \quad \text{where} \quad \omega_0^2 = \frac{g}{\ell}.$$

---

[7] With kind permission of himself and his parents.

Now we consider the coupled situation with spring constant D, see figure above. The horizontal displacement of the right-hand mass is denoted by y.

We wish to establish Newton's law of motion for the two coupled pendulums. In addition to the restoring forces of the weights, we need to incorporate the spring force. In the situation shown in the figure, the spring forces act on the left-hand mass to the right, on the right-hand mass to the left. Therefore,

$$\left.\begin{array}{l} m\ddot{x} = -m\omega_0^2 \cdot x + D \cdot (y-x) \\ m\ddot{y} = -m\omega_0^2 \cdot y - D \cdot (y-x) \end{array}\right\}.$$

Dividing by m and setting $k = \frac{D}{m}$ results in the following system of linear homogeneous second order differential equations:

$$\left.\begin{array}{l} \ddot{x} = -\omega_0^2 \cdot x + k \cdot (y-x) \\ \ddot{y} = -\omega_0^2 \cdot y - k \cdot (y-x) \end{array}\right\}. \tag{4.10}$$

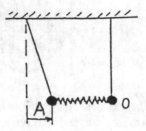

We have to deal with two initial displacements and two initial velocities.

It turns out that for the physical discussion it suffices to look at the situation with the following initial conditions, see figure:

$$x(0) = A, \quad y(0) = \dot{x}(0) = \dot{y}(0) = 0. \tag{4.11}$$

The solution to the IVP (4.10) and (4.11) is the superposition of two oscillations with the angular frequencies ω_0 and $\omega_1 = \sqrt{\omega_0^2 + 2k}$ (verification in Exercise 75):

$$x(t) = \frac{A}{2}\{+\cos(\omega_1 t) + \cos(\omega_0 t)\},$$

$$y(t) = \frac{A}{2}\{-\cos(\omega_1 t) + \cos(\omega_0 t)\}. \tag{4.12}$$

Now if weak damping ($k \ll \omega_0$) is present, the two frequencies $\omega_1 > \omega_0$ are almost the same. In this case, it is appropriate to write the solutions in product form by using the trigonometric identities

$$\cos\alpha + \cos\beta = 2\cos\frac{\alpha-\beta}{2}\cdot\cos\frac{\alpha+\beta}{2}, \quad \cos\alpha - \cos\beta = -2\sin\frac{\alpha-\beta}{2}\cdot\sin\frac{\alpha+\beta}{2},$$

where $\alpha = \omega_1 \cdot t$ and $\beta = \omega_0 \cdot t$:

$$x(t) = A\cos(\frac{\omega_1 - \omega_0}{2}\cdot t)\cdot\cos(\frac{\omega_1 + \omega_0}{2}\cdot t), \tag{4.13}$$

$$y(t) = A\sin(\frac{\omega_1 - \omega_0}{2}\cdot t)\cdot\sin(\frac{\omega_1 + \omega_0}{2}\cdot t). \tag{4.14}$$

The leading factors have a very small frequency $\dfrac{\omega_1 - \omega_0}{2}$ compared to the trailing factors with frequency $\dfrac{\omega_1 + \omega_0}{2} \approx \omega_0$.

The equations (4.13) and (4.14) describe a so-called **amplitude modulation**, as is shown by the following diagram:

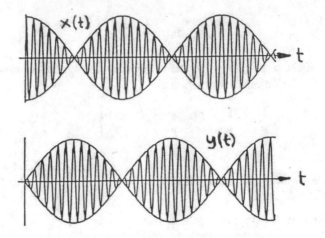

There is an exchange of energy. If one pendulum is practically at rest, then the other is at its peak, and vice versa.

In acoustics this phenomenon is called **beats**.

4.11 Predator-Prey Problem

The **Lotka-Volterra model** was proposed by the chemist Alfred Lotka (1880–1949) and the physicist Vito Volterra (1860–1940). It is of significance in biology and economics.

A typical problem is the **predator-prey interactions** that describes the behavior of two animal species, such as foxes (predators) and hares (prey). It is represented by the following system of nonlinear autonomous differential equations:

$$\left.\begin{array}{l} \dot{h} = a \cdot h \ - b \cdot h \cdot f \\ \dot{f} = -c \cdot f + d \cdot h \cdot f \end{array}\right\} . \tag{4.15}$$

Here $f(t)$ is the time-varying population of the predators and $h(t)$ that of the prey. Equation (4.15) is called a **dynamical system**.
The constant parameters a, b, c, d are all positive.

The four terms on the right have the following meanings:

- $a \cdot h$ is the reproduction rate of the prey with ample food supply and in the absence of predators.
- $-c \cdot f$ is the death rate of the predators in the absence of prey.
- $d \cdot h \cdot f$ is the reproduction rate of the predators without the death rate.
- $-b \cdot h \cdot f$ is the decimation rate of the prey without the reproduction rate.

The product terms $b \cdot h \cdot f$ and $d \cdot h \cdot f$ express the fact that the rates are proportional to the number of encounters between the two species.

The system of differential equations has no elementary solution. For this reason, we shall use numerical methods.

Evidently, this is an idealized model (there is no third species involved, external influences are considered constant).

Lotka and Volterra proved that the **solutions are periodic**, and that the predator population shows a behavior delayed in time: The maximum of the predator population is reached after the maximum of the prey population. Reason: If the number of prey is large, the reproduction rate of the predators is high. But it takes a while, until the baby animals are born. If the predators increase in number, then the prey population shrinks. If the prey population shrinks, then the hunting success of the predators declines, whence the predator population shrinks due to food shortage. This in turn allows the prey population to increase.

The two lines of the Mathematica code
$b = c = d = 1; a = 2;$
StreamPlot[$\{$a h $-$ b h f, $-$c f $+$ d h f$\}, \{h, 0, 5\}, \{f, 0, 5\}$]
containing the right-hand side of the vector field of (4.15) provide us with the following so-called **phase portrait** (where h is the abscissa and f is the ordinate):

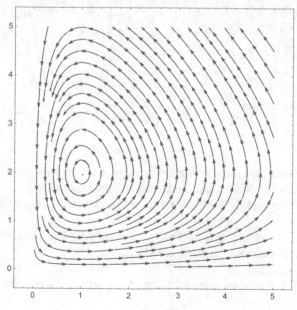

These are parametrized curves that revolve around the **stable equilibrium point**
$(h, f) = (1, 2)$ in counterclockwise direction. The equilibrium point is also called a
stable fixed point, because it remains fixed in time. In its vicinity the curves are
almost ellipses (see Exercise 80).
The constant solution corresponds to the coordinates of the fixed point

$$h = 1, \quad f = 2.$$

In general, (expressed somewhat inaccurately), an **equilibrium point is stable**, if
every solution that has an initial condition close enough to the fixed point stays in
its vicinity. Asymptotically stable is thus a special case of stable.

The following program code numerically solves the system of differential equations
(4.15) for the initial values $h(0) = 3$ and $f(0) = 2$. In addition, it displays the graphs
of the two functions h and f.[8]

```
eq1 = h'[t] == a * h[t] - b * h[t] * f[t];
eq2 = f'[t] == -c * f[t] + d * h[t] * f[t];
eq3 = h[0] == 3;
eq4 = f[0] == 2;
loes = Flatten[NDSolve[{eq1, eq2, eq3, eq4, h[t], f[t]}, {t, 0, 15}]];
{hh[t_], ff[t_]} = {h[t], f[t]} /. loes;
Plot[{hh[t], ff[t]}, {t, 0, 15}]
```

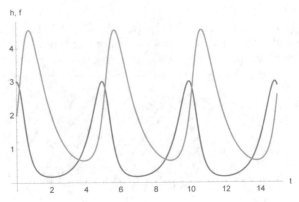

Remarks:

(a) Here the numerical values for the parameters a, b, c, d were not chosen realisti-
    cally, but rather in order to enhance the qualitative behavior.
(b) The model may rightly be criticized. A refined model is suggested in Exercise
    95.
(c) Another important finding is the fact that not only are the solutions periodic, but
    the **average population sizes are constant**.

---

[8] The book [37] features many examples that use the software package *Mathematica* to solve
problems.

Example 4.40

According to the trading records of the Hudson's Bay Company that date back to around 1845–1930, the number of pelts of lynx (predators) and of snow hares (prey) are periodic with a period of 9.6 years. Admittedly, this system was interfered by a second species of predators: the hunters of the Hudson's Bay Company.

In this example, there was an extreme fluctuation of the annual number of pelts of lynx, namely between 1,000 and 70,000. Similarly, the annual number of pelts of snow hares fluctuated extremely between 2,000 and 160,000. But the average over several periods of 9.6 years remained at about 20,000 lynx pelts and 80,000 snow hare pelts annually. ◇

4.12 Periodic Solutions and Limit Cycles

The theoretical foundation is provided by the theorem of Poincaré[9] and Bendixson.[10] It states that for a certain class of systems of autonomous differential equations, there are periodic solutions or limit cycles that are solutions. This is an existence theorem that says nothing about how to calculate the solutions. We shall not go into the details, but refer to [4].

The predator-prey problem belongs to the category of problems with periodic solutions.

In exploring limit cycles we consider the following system of differential equations in the time-varying Cartesian coordinates $x(t)$, $y(t)$:

$$\dot{x} = x - y - x \cdot \sqrt{x^2 + y^2} \,,$$
$$\dot{y} = x + y - y \cdot \sqrt{x^2 + y^2} \,.$$

This problem may be solved analytically by means of polar coordinates (r, φ):

$$x\dot{x} + y\dot{y} = x^2 + y^2 - (x^2 + y^2)\sqrt{x^2 + y^2} = r^2 - r^3 \,. \tag{4.16}$$

Moreover,

$$\frac{d}{dt}(x^2 + y^2) = 2x\dot{x} + 2y\dot{y} = \frac{d}{dt}(r^2) = 2r\dot{r} \,.$$

Substitution into (4.16) results in

$$r\dot{r} = r^2 - r^3 \quad \Longrightarrow \quad \dot{r} = r(1 - r) \,.$$

On the right-hand side we have a logistic differential equation with $a = b = 1$, however, in the polar coordinate r. The general solution according to Exercise 66 (b) is

---

[9] Henry Poincaré (1854 – 1912) was a distinguished French mathematician, physicist, astronomer and philosopher. The Poincaré Conjecture was one of the most significant unsolved problems in mathematics. For its solution a prize of one million dollars was offered. The Russian mathematician Grigori Perelman (born 1966) solved this millenium problem in 2002. He rejected both the prize money and the Fields Medal (the latter corresponding to the Nobel Prize).

[10] The Swedish mathematician Ivar Bendixon (1861–1935) proved the theorem in a more general context than Poincaré. The theorem says that periodic solutions or limit cycles exist as solutions for a certain class of autonomous differential equation systems. It is an existence theorem that does not make any statement as to how such solutions can be calculated. Further details are in [4].

$$r(t) = \frac{1}{1 + De^{-t}} \quad \text{where} \quad D > -1 .$$

Now we need to determine a relationship between the time t and the angle φ. As before, we obtain

$$y\dot{x} - x\dot{y} = -x^2 - y^2 = -r^2 . \tag{4.17}$$

With the abbreviations $x = r \cdot \cos\varphi = r \cdot c$ and $y = r \cdot \sin\varphi = r \cdot s$ where both r and φ depend on t, we obtain

$$y\dot{x} - x\dot{y} = sr(c\dot{r} - s\dot{\varphi}r) - cr(\dot{\varphi}cr + s\dot{r}) = -(c^2 + s^2)\dot{\varphi}r^2 = -\dot{\varphi}r^2 .$$

Substitution into (4.17) yields $-\dot{\varphi}r^2 = -r^2 \implies \dot{\varphi} = 1 \implies \varphi(t) = t + \varphi_0$.
With the initial values $r(0) = r_0$ and $\varphi(0) = \varphi_0$ we obtain $D = \frac{1-r_0}{r_0}$. Thus

$$r(\varphi - \varphi_0) = \frac{1}{1 + \frac{1-r_0}{r_0}e^{-(\varphi - \varphi_0)}} \quad \text{where} \quad \varphi \geq \varphi_0 .$$

The graphs inside ($0 < r_0 < 1$) and outside ($r_0 > 1$) the unit circle spiral to it asymptotically in counterclockwise direction. The unit circle is a periodic solution ($r_0 = 1$) and is called a **stable limit cycle**.

The figure displays a rotational symmetry for r_0 fixed and φ_0 varying at steps of $\frac{\pi}{2}$. Obviously, the origin is an **unstable equilibrium point**.

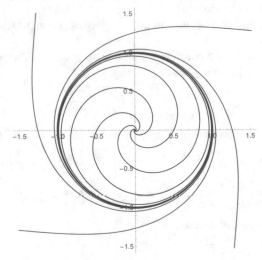

The representation of the solutions in the (x, y)-plane is called a **phase portrait**.

One application is the so-called brusselator,[11] a model for chemical oscillators, such as the Belousov-Zhabotinsky reaction. This is described by four reaction equations where the concentrations of the two substances show a time-periodic behavior. See, for instance, [48] on the Internet. The name brusselator is a concatenation of the words "Brussels", where the theory was developed, and "oscillator".

---

[11] In this context, the Nobel prize laureate for Chemistry, Ilya Prigogine (1917–2003), a Russian-Belgian physicist, chemist and philosopher, must be named: He did significant work on dissipative structures, self-organisation and irreversibility. In [11] the topic of bifurcation of a brusselator with symmetric building structures is discussed.

4.13 Two-Body Problem in Celestial Mechanics

To conclude this chapter, we look at a problem that has kept humankind busy for centuries.

4.13.1 Historical Remarks

The Danish astronomer[12] Tycho Brahe (1546–1601), who as an aristocrat largely financed his research himself, set up two well-equipped observatories, "Uraniborg" and later "Stjerneborg", on the island of Hven. This island was given to him by the Danish king. There he spent more than 20 years conducting observations and measurements with improved and newly invented instruments.

His observations were of highest accuracy ever achieved before the invention of the telescope. The errors in angle measurements were at instances as low as 30 seconds, i.e. about $0.01°$.

His data included over 1000 fixed star positions, the discovery of irregularities in the lunar motion, as well as comprehensive and accurate measurements of planetary motion.

In 1597 Brahe left his homeland and after a few stops the German emperor Rudolf II granted him a position and a castle for his work near Prague. For evaluating his observation material, Brahe hired two assistants, the Dane Longomontanus and in 1600 Kepler (1571–1630).

Brahe did not consider his observational activities his main achievement, but rather his establishing a new concept of the world: Although he adhered to the idea of the earth-centered world by letting the sun revolve around the earth, he incorporated the approach of the heliocentric world view insofar as he moved the sun into the center of all the other planetary orbits.

Even on his deathbed, he implored Kepler to fight for the Tychonian world view. But Brahe's observation data convinced Kepler to reject all forms of geocentric models.

When Kepler began his work with Brahe, he already had a name as a mathematician and astronomer himself: In 1596, his "Mystery of the Cosmos" (Mysterium Cosmographicum) appeared, a fascinating mixture of speculation and science. The six spheres, on which the then known planets revolved with the sun in the center, were associated with the five Platonic solids (regular tetrahedron, cube, octahedron, icosahedron, dodecahedron).

He received enthusiastic approval from Galileo Galilei (1564–1643) of Padua, but critical hints from Brahe.

Under the pressure of Brahe's data, Kepler revised his idea of circular orbits. Finally he postulated the three famous laws named after him.

---

[12] From [38].

Kepler's Laws:

1. All planets move about the sun in elliptical orbits, with the sun in one of the foci.

2. A radius vector joining any planet to the sun sweeps out equal areas in equal times (law of equal areas).

3. The square T^2 of the orbital period of a planet about the sun is proportional to the cube a^3 of the semimajor axis of its ellipse: Or equivalently, for two planets 1 and 2 the following is true:

$$\frac{T_1^2}{T_2^2} = \frac{a_1^3}{a_2^3} \,.$$

The third law was found much later than the first two, namely on May 18, 1618, five days prior to the Second Defenestration of Prague, marking the beginning of the Thirty Years' War.

The following diagram[13] illustrates Kepler's second law:

Neighboring points on the ellipse have a constant time difference Δt and all triangles (with a curved side) have the same constant area.

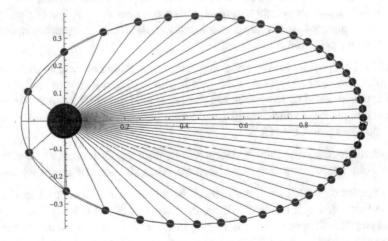

Only 13 years after Kepler's death, Newton (1643–1727) was born. Kepler's laws brought him to the conclusion that gravitation exists and he taught how to master its computation while establishing his celestial mechanics. On his way, he developed the infinitesimal calculus.

---

[13] From a project by my former students Amrein and Graf, carried out in a mathematics workshop of the Department of Technology and Computer Science at the Bern University of Applied Sciences.

4.13.2 Proof of Kepler's Second Law

The following calculations are valid for a closed two-body system, that is, if there are no disturbing forces, such as by another celestial body.

This applies well to

- sun and planet
- earth and satellite

Reason: Due to mass and distance, other celestial bodies have no relevance.

The calculations are done in polar coordinates with the sun at the origin of the reference frame.

The time-varying velocity vector is also described in polar coordinates:

- radial component: $v_r = \dot{r}(t)$.
- tangential component: $v_\varphi = r \cdot \dot{\varphi}$.

The angular momentum $\vec{D}$ abides to the principle of conservation:

$$\vec{D} = \vec{r} \times m\dot{\vec{r}} = \overrightarrow{\text{const.}}$$

Hence velocity and position vectors are always in the same plane, implying a **plane motion**.

The magnitude D of the angular momentum is

$$D - mr^2 \dot{\varphi}(t) - \text{const.} \tag{4.18}$$

According to the figure on the right, where $\Delta\varphi$ is small, the increment of the area is

$$\Delta A \approx \frac{1}{2} r \cdot r \cdot \Delta\varphi \Longrightarrow \frac{\Delta A}{\Delta t} = \frac{1}{2} r^2 \frac{\Delta\varphi}{\Delta t} = \frac{D}{2m} = \text{const.}$$

If we take the limit and use Equation (4.18), we obtain

$$\frac{dA}{dt} = \frac{1}{2} r^2 \dot{\varphi} = \frac{D}{2m} = \text{const.}$$

This relation expresses Kepler's Second Law mathematically, because it implies that $A(t)$ is a **linear function**. Hence ΔA is the same for every finite time interval of fixed length Δt.

4.13.3 Proof of Kepler's First Law

The dominant mass of the two-body system is regarded as at rest. Thus we study the motion of the second body relative to the dominant body.

The principle of energy conservation states: The sum of the kinetic energy T and the potential energy V is constant at all times, if there is no frictional loss. This assumption is practically satisfied for the two-body problem.

$$T + V = \frac{1}{2}mv^2 - G\frac{M \cdot m}{r} = \text{const} = E = \text{total energy}.$$

Here $G = 6.6725985 \cdot 10^{-11} \text{Nm}^2\text{kg}^{-2}$ is the universal gravitational constant. It is one of the most accurately measurable constants of the natural sciences.

We express v^2 in polar coordinates:

$$v^2 = v_r^2 + v_\varphi^2 = \dot{r}^2 + (r\dot{\varphi})^2 = (\frac{dr}{d\varphi} \cdot \dot{\varphi})^2 + r^2\dot{\varphi}^2 = \dot{\varphi}^2[(\frac{dr}{d\varphi})^2 + r^2] = \left(\frac{D}{mr^2}\right)^2 (r'^2 + r^2).$$

The last equality follows from (4.18), where the following short notations are used: $' = \frac{d}{d\varphi}$ and $\dot{} = \frac{d}{dt}$.

Next we substitute the last result of v^2 into the energy equation above:

$$\frac{1}{2}m \cdot \frac{D^2}{m^2 r^4}[r'^2 + r^2] - G\frac{mM}{r} = E.$$

Then we multiply by $2m/D^2$:

$$\frac{1}{r^4}r'^2 + \frac{1}{r^2} - 2G\frac{m^2 M}{D^2} \cdot \frac{1}{r} = \frac{2mE}{D^2}. \tag{4.19}$$

Substituting the reciprocal $u(\varphi) = 1/r(\varphi)$ and differentiation implies

$$\frac{du}{d\varphi} = -\frac{1}{r(\varphi)^2} \cdot \frac{dr}{d\varphi} \implies (\frac{du}{d\varphi})^2 = \frac{1}{r(\varphi)^4} \cdot (\frac{dr}{d\varphi})^2.$$

Substituted into (4.19) gives

$$(\frac{du}{d\varphi})^2 + u(\varphi)^2 - 2\left(\frac{Gm^2 M}{D^2}\right)u(\varphi) = \frac{2mE}{D^2}.$$

The positive constant between the set of large parentheses is denoted by $1/p$ and the positive constant on the right by λ, thus leading to the separable differential equation

$$\frac{du}{d\varphi} = \sqrt{\lambda - u(\varphi)^2 + \frac{2}{p} \cdot u(\varphi)}.$$

Therefore,

$$\int \frac{du}{\sqrt{\lambda - u(\varphi)^2 + \frac{2}{p} \cdot u(\varphi)}} = \int d\varphi + C.$$

The integration is a bit tricky (Mathematica cannot do it):

$$\varphi = \arccos\left(\frac{1/p - u}{\sqrt{1/p^2 + \lambda}}\right) + C.$$

Remark: Verify by differentiating.

As we are only interested in the orbit, we may set the beginning of the angle measurement arbitrarily. Therefore, we take the integration constant $C = 0$ and solve for u step by step:

$$\cos \varphi = \frac{1/p - u}{\sqrt{1/p^2 + \lambda}} \implies \cos \varphi \cdot \sqrt{1/p^2 + \lambda} = 1/p - u.$$

Hence

$$u = 1/p - \cos \varphi \sqrt{1/p^2 + \lambda} = \frac{1}{p}[1 - \cos \varphi \cdot \sqrt{1 + \lambda p^2}].$$

The radical is a constant and after resubstituting p and λ, it equals

$$\varepsilon = \sqrt{1 + \lambda p^2} = \sqrt{1 + \frac{2D^2}{G^2 m^3 M^2} E}.$$

With $r = 1/u$, we obtain a conical section

$$r(\varphi) = \frac{p}{1 - \varepsilon \cos \varphi}.$$

The following holds:

$$\varepsilon \begin{cases} < 1: & \text{ellipse} \\ = 1: & \text{parabola} \\ > 1: & \text{hyperbola} \end{cases} \iff \text{total energy} \quad E \begin{cases} < 0 \\ = 0 \\ > 0 \end{cases}$$

where **origin = focus**. This proves Kepler's First Law (and more).

Remark: The polar equations of the remaining conical sections, the hyperbola and the parabola, may be obtained similarly.
If $\varepsilon \geq 1$, then evidently r grows infinitely.

4.13.4 Proof of Kepler's Third Law

The numerical eccentricity was defined as

$$\varepsilon = \frac{e}{a} = \frac{\sqrt{a^2 - b^2}}{a} .$$

Hence $\varepsilon^2 = 1 - \dfrac{b^2}{a^2} \implies \dfrac{b^2}{a^2} = (1 - \varepsilon^2) \implies b^2 = a^2(1 - \varepsilon^2) ,$

which we shall use in the following:

$$p = \frac{b^2}{a} = a(1 - \varepsilon^2) = \frac{D^2}{m^2 MG} .$$

The second equality for p was defined earlier.

Due to $\dot{A} = \frac{1}{2} r^2 \dot{\varphi}(t) = \text{const}$ and conservation of angular momentum, we obtain $D = mr^2 \dot{\varphi}(t) = \text{const}$, and $D = 2m\dot{A}$. Substituting this into the above equation for p yields

$$p = a(1 - \varepsilon^2) = \frac{4m^2 \dot{A}^2}{m^2 MG} .$$

Canceling m^2 and solving for $\dot{A}$ results in

$$\dot{A} = \frac{1}{2} \sqrt{aMG(1 - \varepsilon^2)} .$$

In order to compute the orbital period T, we substitute this result into the following expression:

$$T = \frac{\text{area of ellipse}}{\dot{A}} = \frac{\pi ab}{\dot{A}} = \frac{\pi a^2 \sqrt{1 - \varepsilon^2}}{\dot{A}} = \frac{2\pi a^2}{\sqrt{aMG}} \frac{2\pi}{\sqrt{MG}} \cdot a^{3/2}.$$

This proves Kepler's Third Law:

$$\frac{T^2}{a^3} = \frac{4\pi^2}{MG} = \text{const.}$$

Or alternatively, if T_i are the orbital periods and a_i the semimajor axes of two planetary orbits with $i = 1, 2$, then

$$\frac{T_1^2}{T_2^2} = \frac{a_1^3}{a_2^3} .$$

Example 4.41
From the orbital period $T_{\text{Pluto}} = 247.7$ years of Pluto (which in 2006 was downgraded to a dwarf planet) and the earth's semimajor axis $a_{\text{Earth}} = 149.6 \cdot 10^6$ km, we may compute the semimajor axis of Pluto's orbit:

$$247.7^2 = \frac{a_{\text{Pluto}}^3}{a_{\text{Earth}}^3} \implies a_{\text{Pluto}} = 247.7^{(2/3)} a_{\text{Earth}} = 5910 \cdot 10^6 \text{ km.} \qquad \Diamond$$

4.14 Exercises Chapter 4

72. Mechanical Oscillator. Given a free undamped mechanical oscillator, compute the general solution and the amplitude A of the Initial value problem

$$\ddot{y} + \omega^2 y = 0,$$

where $y(0) = y_0$ is the initial displacement and $\dot{y}(0) = v_0$ the initial velocity.

73. Toy Oscillator. A weak spring with spring constant D together with an object of mass m, for example a ball or a small doll, are given. If the object is attached to the spring and displaced from its equilibrium level, then the object oscillates up and down.

In a numerical example, the mass is $m = 80$ g and the time for an up-and-down movement is measured to be $T = 0.90$ s. Find a differential equation and compute D.

74. A Child's Dream. Consider a full sphere of mass M with uniform density and a point mass m inside the sphere at distance r from its center. Interestingly, mass m is attracted to the sphere's center by a force that depends **linearly** on r.[14]

Now let us assume that the earth has radius $R = 6380$ km, is of uniform density, cold in the interior, and that a hole is drilled straight through its center.

What happens if you fall through the hole? Show that in the absence of friction you would oscillate harmonically between the two ends of the hole.

Also compute the angular frequency ω, the oscillation period T, and the maximum velocity v_{max} of your motion.

75. Weakly Damped Oscillator and Coupled Pendulums. Verify the solutions

(a) in the case of the free weakly damped oscillator (4.7) with initial conditions $y(0) = y_0$, $\dot{y}(0) = v_0$.

(b) in the case of the coupled pendulums (4.12) with initial conditions (4.11).

76. Breakage of a Spring. A mass of 10 kg attached to a spring stretches the latter by 10 cm. If the mass oscillates freely, then the amplitude decreases per oscillation period by 3%. It is known that the spring breaks, if the amplitude exceeds 8 cm. Question: Does the spring break, if it is incited by a periodical force $F = \cos(\Omega t)$ N, i.e. with an amplitude of 1 N?

Hint: First compute the spring constant D and the oscillation period T.

---

[14] For a mass m outside the mass M at distance r, the gravitational force is known to be proportional to $1/r^2$.

77. Festival of Hyperbolic Functions.

A completely flexible rope or a smooth ribbon of length ℓ and mass m slides friction-
lessly over the edge of a table, see figure. At
time $t = 0$ the rope or ribbon is released at
height $h(0) = h_0$.

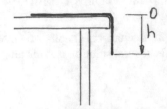

(a) Find the equation of motion for the height $h(t)$, its solution and the velocity
    function $v(t)$.
(b) Compute the time t_{end} when the rope or ribbon leaves the table.
(c) For $\ell = 1.5$ m compute the time t_{end} and the velocity $v(t_{end})$ for each of the two
    initial conditions $h_0 = 0.5$ m and $h_0 = 0.05$ m.

78. Magnetic Field of a Straight Conductor, Biot-Savart's Law.

An infinitely long straight wire ℓ carrying a current i generates an $\vec{H}$-field with axial symmetry about the wire, see figure. Biot-Savart's Law
states $\vec{dH} = \frac{i}{4\pi} \frac{\vec{d\ell} \times \vec{r}}{r^3}$. The vector $\vec{dH}$ is perpendicular to the plane containing the conductor
and the point P. The field lines are concentric circles of radius a around the conductor.
Compute the magnitude of the magnetic field
$H = |H(P)|$ by integrating with respect to φ.

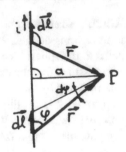

79. RC Circuit. At time $t = 0$ the initially uncharged capacitor is connected to a
constant input voltage U. Compute the charge function $q(t)$ as well as the current
function $i(t)$ and sketch their graphs.

80. Predator-Prey Model. Let us consider the system of nonlinear differential
equations (4.15). We wish to find an approximate solution in a small neighborhood
of the stable equilibrium point $P(h_0, f_0)$.

Proceed as follows:

(a) Compute h_0, f_0 from the given positive parameters a, b, c, d.
(b) Introduce a new shifted coordinate system $(x, y) = (h - h_0, f - f_0)$ with origin at
    P. Rewrite the system of equations in terms of the new variables x, y and linearize
    the system by discarding the product terms $x \cdot y$; they are negligible, because their
    absolute values are small compared to the linear terms.
(c) Use $\frac{dy/dt}{dx/dt} = y'(x)$ and show that the solutions of the differential equation for
    $y(x)$ are approximated by similar ellipses centered at P. The ratio between their
    semimajor axes is $\left(\frac{b}{d}\sqrt{\frac{c}{a}}\right) : 1$.

81. Coriolis Force in the Southern Hemisphere. Solve Problem (4.3), (4.2) for $f < 0$ in the southern hemisphere and give a geometric interpretation.

82. Initial Value Problem. A mass of weight 19.6 N extends a spring by 4 cm. The mass is displaced 5 cm from its equilibrium position. An external periodic force $F = 10\cos(3t)$ N is imposed on it practically undamped.
Determine an IVP and specify the dimensions.

83. Weak Damping. Consider the IVP

$$\ddot{y} + 0.125\dot{y} + y = 3\cos(\Omega t).$$

(a) Give reasons that this is an oscillation phenomenon with weak damping.

(b) Compute the resonance amplitude and the resonance frequency.

84. System of Differential Equations. Given the initial value problem

$$\left. \begin{array}{rcl} x'(t) &=& 2x(t) - 3y(t) \\ y'(t) &=& y(t) - 2x(t) \\ x(0) &=& 8 \\ y(0) &=& 3 \end{array} \right\}.$$

Verify that the solution is $x(t) = 5e^{-t} + 3e^{4t}$, $y(t) = 5e^{-t} - 2e^{4t}$.

85. Perfect Resonance. Given the initial value problem

$$\left. \begin{array}{rcl} \ddot{y}(t) + \omega^2 y(t) &=& A\cos\Omega t \\ y(0) &=& \alpha \\ \dot{y}(0) &=& \beta \end{array} \right\}.$$

(a) Verify that the solution is

$$y(t) = \frac{A(\cos\Omega t - \cos\omega t)}{\omega^2 - \Omega^2} + \alpha\cos\omega t + \frac{\beta}{\omega}\sin\omega t.$$

(b) Give a physical interpretation of ω.

(c) Discuss the behavior of the solution in the special case $\alpha = \beta = 0$ for $\Omega \approx \omega$.
Between what values does y oscillate?

86. Beats. Rewrite the function

$$f(t) = \cos(17t) - \cos(19t)$$

as a product of two trigonometric functions and sketch by hand its graph on the interval $0 \le t \le 4\pi$.

87. LCR Circuit with Input Voltage. At time $t = 0$ the circuit with input voltage $E(t)$ and an uncharged capacitor with $q(0) = 0$ is closed.

(a) Formulate an IVP with the parameters $L = 2$ H, $C = 0.02$ F, $R = 16$ Ω for the charge function $q(t)$.

(b) Given a constant input voltage of $E(t) = 300$ Volt, verify by substitution that

$$q(t) = 6 - e^{-4t}(6\cos 3t + 8\sin 3t)$$

is the solution and compute the current function $i(t)$.

(c) Next the input is an alternating current voltage of $E(t) = 100 \cdot \sin(3t)$ Volt. Establish a differential equation for the current function $i(t)$ by differentiating the one for the charge function obtained in (a). Verify that

$$i(t) = \frac{75}{52}(2\cos 3t + 3\sin 3t) - \frac{25}{52}e^{-4t}(17\sin 3t + 6\cos 3t)$$

is the solution and determine the prevailing steady-state current in the form

$$i_{st}(t) = A\sin(\omega t + \varphi),$$

once the transient component has died down.

88. Planetary Orbits. Prove Kepler's Third Law $\dfrac{T^2}{r^3} = $ const in the case of circular planetary orbits with radius r and orbital period T by use of the central force $m\dfrac{v^2}{r}$ and the gravitational force $F = G\dfrac{mM}{r^2}$.

Chapter 5
Numerical Methods With Applications

5.1 Basics

Numerical methods allow us to solve almost "arbitrarily complicated" ordinary differential equations[1] and to graph the solution curves. They have the great advantage that they do not depend on whether or not integration is elementary. This is undoubtedly a gain in flexibility over analytical methods. Moreover, it is crucial for practical problems, which are often more complicated than textbook examples.

However, there are two major disadvantages of the numerical over the analytical methods:

- Numerical methods do not allow parameters as symbols. This eliminates the immediate interpretation of a solution in terms of parameters.

- Numerical methods are always approximation methods. Consequently, it must be ensured that the errors are kept small, so that the numerical approximation yields a useful solution. Such an endeavor can be problematic without much experience and knowledge of the respective algorithms[2]!

The first problem can be somewhat mitigated. We solve the same differential equation multiple times by selecting a series of different numerical parameter values and compare their graphs. However, if two or even more parameters are involved, a useful discussion of the solutions becomes difficult or even impossible.

---

[1] For partial differential equations involving functions of several independent variables, there are so-called discretization methods among the numerical methods, see, for example, [11].

[2] See example under *Pitfalls* in [37].

© The Author(s), under exclusive license to Springer Nature Switzerland AG 2021
A. Fässler, *Fast Track to Differential Equations*,
https://doi.org/10.1007/978-3-030-83450-0_5

5.2 Euler's Method

Let us begin with the general first order differential equation with one initial condition:

$$y' = f(x,y), \quad y(x_0) = y_0.$$

In this chapter, we usually choose x as the independent variable. The given function $f(x,y)$ is piecewise continuous in the two independent variables x and y, but otherwise "arbitrarily complicated". The given point (x_0, y_0) corresponds to the initial condition of the desired solution $y(x)$.

Euler's idea is simple: We start at point (x_0, y_0) and proceed with a small step size $\Delta x = h$ in direction of the tangent with the known slope $m = f(x_0, y_0)$ to obtain an approximate point (x_1, y_1) of the solution curve.
This gives rise to

$$y_1 = y_0 + h \cdot f(x_0, y_0)$$

$$x_1 = x_0 + h.$$

Now we iterate this process with point (x_1, y_1) as the new starting point to obtain point (x_2, y_2), etc. Hence **Euler's method** is simply given by

$$y_{k+1} = y_k + h \cdot f(x_k, y_k)$$

$$x_{k+1} = x_k + h$$

where $k = 0, 1, 2, \ldots, n-1$.

The quantities $y_k = y(x_k)$ describe approximate values of the solution. The linear interpolation between the individual points produces a polygonal graph as an approximate solution curve.

Under the idealized assumption that the computer can compute real numbers accurately (i.e. with infinitely many digits), we may expect that in every numerical method with step size h the calculated values approach the exact solution as $h \to 0$.

Let us test Euler's method in this sense by way of the model problem

$$y' = y \quad \text{where} \quad y(0) = 1.$$

It has the exact solution $y(x) = e^x$. After n steps, Euler's method as described above yields

$$y_1 = 1+h$$
$$y_2 = y_1 + y_1 \cdot h = y_1(1+h) = (1+h)^2$$
$$y_3 = y_2 + y_2 \cdot h = y_2(1+h) = (1+h)^3$$
$$\ldots$$
$$y_n = (1+h)^n$$

Now we consider a **fixed location** $x = n \cdot h$ and take the limit of y_n as $n \to \infty$:

$$\lim_{n\to\infty} \left(1+\frac{x}{n}\right)^n = e^x .$$

As a matter of fact, we do obtain the exact solution of the initial value problem.

Analytical investigations show that Euler's method converges towards the exact solution not only in our simple model problem, but also in general.

5.3 Error Considerations

Now it doesn't make sense to argue that we should just let the step size h go to 0 in order to get an exact solution. The reason is that it would be time-consuming and the computational effort would go to ∞!

In addition to the aforementioned **methodical errors M**, we have **roundoff errors R** when working with finitely many digits. Unfortunately, these errors can accumulate to an alarming size.

The smaller the step size, the larger the accumulated roundoff error! Therefore, the following qualitative situation may occur, in which the two error sources realistically must be added up:

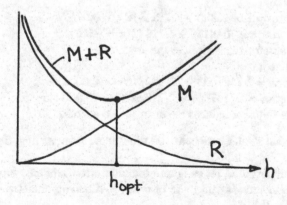

From the diagram we infer that there is an optimal step size h_{opt}. Unfortunately, it depends not only on the numerical method chosen, but even more so on the differential equation itself: It is obvious that if a solution shows rapid oscillations, then the step size should be chosen small. Otherwise the methodical error could be massive.

However, it is impossible to analytically estimate the roundoff error in the general case. Therefore, it is vital to gather experience with numerical examples that allow an exact analytical solution.

5.4 Heun's Method

Heun's idea:[3] After calculating an initial Euler step, one takes the arithmetic mean of the two slopes for the final Euler step, as illustrated in the figure:

$$Y = y_k + h \cdot f(x_k, y_k)$$

$$y_{k+1} = y_k + \frac{h}{2} \cdot [\underbrace{f(x_k, y_k)}_{m_1} + \underbrace{f(x_{k+1}, Y)}_{m_2}]$$

where $k = 0, 1, 2, \ldots$

Example 5.42 Let us reconsider our model problem

$$y' = y \text{ where } y(0) = 1.$$

Use both methods to calculate the function value $y(1)$, first with 50 steps ($h = 0.02$), then with 100 steps ($h = 0.01$), and give the difference to the true solution, i.e. the error $y(1) - e$:

- Euler's method:
  50 steps: error $= 2.6915880 - 2.7182818 = -0.0267$
  100 steps: error $= 2.7048138 - 2.7182818 = -0.0135$
  Halving the step size also halves the error.
- Heun's method:
  50 steps: error $= 2.7181033 - 2.7182818 = -0.000179$
  100 steps: error $= 2.7182369 - 2.7182818 = -0.000045$
  Halving the step size reduces the error to one quarter.

In addition, Heun's values are over 100 times more accurate than Euler's values! ◇

For our present purposes, let us ignore roundoff errors and only look at methodical errors. Then a method is said to be of order p, if halving the step size reduces the error to a factor of 2^{-p}.

The methodical error refers to small step sizes h. The smaller h, the better does the ratio of the calculated errors approach 2^{-p} when halving the step size.

---

[3] Karl Heun (1859–1929) studied mathematics and physics in Göttingen and then taught at an agricultural school.

What we found in the model example by experiment holds in general:

- Euler's method is of order 1.
- Heun's method is of order 2.

5.5 Runge-Kutta Method

The Runge-Kutta method is a further improvement of Euler's method and is often implemented in numerical algorithms.

Here we reproduce the algorithm for the **fourth order Runge-Kutta method**[4] without giving a detailed derivation (which takes a lot of effort): Starting from point (x_k, y_k), we obtain the next point (x_{k+1}, y_{k+1}) by calculating the following steps line by line:

$$k_1 = f(x_k, y_k) \qquad\qquad y_a = y_k + \tfrac{h}{2}k_1$$
$$k_2 = f(x_k + \tfrac{h}{2}, y_a) \qquad\qquad y_b = y_k + \tfrac{h}{2}k_2$$
$$k_3 = f(x_k + \tfrac{h}{2}, y_b) \qquad\qquad y_c = y_k + hk_3$$
$$k_4 = f(x_k + h, y_c)$$

$$y_{k+1} = y_k + \tfrac{h}{3}\left(\tfrac{1}{2}k_1 + k_2 + k_3 + \tfrac{1}{2}k_4\right)$$
$$x_{k+1} = x_k + h$$

Therefore, one Runge-Kutta step consists of four Euler steps.

To better understand this fourth order method described above, we may illustrate the individual values in the following diagram:

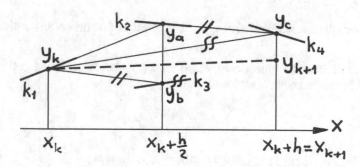

The k_i-values are slopes of the direction field at the four points.
We start from the known point (x_k, y_k). As a preliminary to obtaining the next point (x_{k+1}, y_{k+1}) we calculate three auxiliary points, each by an Euler step starting from (x_k, y_k).

---

[4] Carl Runge (1856–1927) and Wilhelm Kutta (1867–1944) were German mathematicians.

Finally we obtain the new point (x_{k+1}, y_{k+1}) by a last Euler step using the weighted average of the four slopes k_1, k_2, k_3, k_4 with weights $\frac{1}{6}, \frac{1}{3}, \frac{1}{3}, \frac{1}{6}$.[5]

Example 5.43 For the model example $y' = y$ with $y(0) = 1$, the Runge-Kutta method furnishes the following errors $y(1) - e$:

- 50 steps: $2.71827974 - e = -2.088 \cdot 10^{-6}$
- 100 steps: $2.71828169 - e = -1.385 \cdot 10^{-7}$

Comparing the errors: $\dfrac{1.3845 \cdot 10^{-7}}{2.088 \cdot 10^{-6}} = \dfrac{1}{15.08}$, which is roughly $\dfrac{1}{2^4} = \dfrac{1}{16}$.

The errors differ by a factor of 15.08.

Regardless of the differential equation, one can prove analytically that as $h \to 0$ the factor approaches 16. Hence the Runge-Kutta method has an error of order 4.

As one Runge-Kutta step corresponds to about four Euler steps, the computational complexity is about 40 or 80 Euler steps, respectively.

But most important of all, the errors of the Runge-Kutta method are about 100,000 times smaller than the errors of Euler's method, and they are about $\sqrt{100,000} \approx 333$ times smaller than those of Heun's method! $\diamondsuit$

Some algorithms work adaptively with an automatic step-size control: In neighborhoods with large variations they choose a smaller step size than at points where the variation is small.

5.6 Numerical Considerations on Systems of Differential Equations

The methods discussed above also apply to systems of first order differential equations.

To understand how the numerical method works, let us consider the case of two equations in two unknown functions x and y:

$$\left. \begin{array}{l} \dot{x} = f(t, x, y) \\ \dot{y} = g(t, x, y) \end{array} \right\}$$

where $x(0) = a, \quad y(0) = b$.

The issue is to compute discrete function values of the two functions given some chosen step size h.

With the knowledge of the two initial values and the slopes at those points, the function values x_1 and y_1 at $t = h$ can be computed.

---

[5] In [23] one finds various algorithms, one of which is the Runge-Kutta method.

For the second iteration, the two slopes at $t = h$ are now known and the function values x_2 and y_2 at $t = 2h$ can be computed, and so forth.

The following diagram clarifies this:

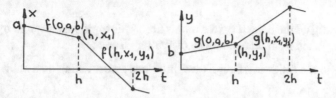

The method works in the same way for n unknown functions.

As we learnt earlier, a differential equation of order n can be rewritten as a system of first order differential equations in n unknown functions. Consequently, higher order differential equations can also be numerically solved by the Runge-Kutta method.

Example 5.44 Lorenz Attractor.

A legendary model is that of meteorologist Edward N. Lorenz (1917–2008),
who brought significant insights to the
theory of **deterministic chaos**. It is described by the following system of nonlinear differential equations:

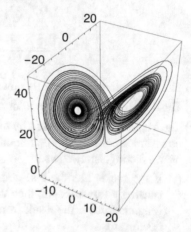

$$\left. \begin{aligned} \dot{x} &= a(y-x) \\ \dot{y} &= x(b-z)-y \\ \dot{z} &= xy - cz \end{aligned} \right\} .$$

The figure[6] shows the associated graph of the numerical solution for parameter values $a = 10$, $b = 28$, $c = 8/3$ with initial values $(x(0), y(0), z(0)) = (0, 1, 0)$.

Using his model, Lorenz discovered that very small changes in the initial conditions can eventually evolve into considerably different solutions. This explains why
precise long-range weather forecasting is impossible. But he also found that the
solutions all are confined to a narrow region of space, hence a certain structure is
maintained. More on this can be found in [2].   ◇

---

[6] Generated by use of the Mathematica commands `NDSolve` and `ParametricPlot3D`.

5.7 Trajectories of Tennis Balls

5.7.1 Modeling

Let us investigate the trajectory of a tennis ball. This topic was inspired by [20] and the data are taken from there courtesy of Walter Gander.

According to the following figure, there are three forces acting on a ball:

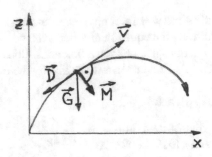

(a) the weight force $\vec{G}$ vertically downwards,

(b) the drag force $\vec{D}$, which opposes the velocity $\vec{v}$,

(c) the Magnus force $\vec{M}$, caused by the spin of the ball. Here the angular velocity $\vec{\omega}$ is assumed constant. By definition the Magnus force is orthogonal to both $\vec{\omega}$ and $\vec{v}$. We distinguish two cases:

  (i) If the racket hits the ball almost tangentially with an upward motion, thus causing a forward spin, then the ball is said to have topspin. In this case the Magnus force is the result from excess pressure above the ball, and $\vec{M}$ points downwards.

  (ii) If the racket hits the ball almost tangentially with a downward motion, thus causing a backward spin, then the ball is said to have slice spin. In this case the Magnus force is the result from excess pressure below the ball, and $\vec{M}$ points upwards.

For the magnitudes of (b) and (c), we have

$$D(v) = C_D \frac{1}{2} \frac{\pi d^2}{4} \rho v^2 \qquad M(v) = C_M \frac{1}{2} \frac{\pi d^2}{4} \rho v^2$$

where $\quad C_D = 0.508 + \left[\dfrac{1}{22.053 + 4.196 \cdot (v/w)^{2.5}} \right]^{0.4} \qquad C_M = \dfrac{1}{2.022 + 0.981 \cdot v/w}$.

Here ρ is the air density and $w = \frac{d}{2} \cdot \omega$ is the circumferential speed, assumed constant.

The coefficients C_D and C_M were determined in [20] by experiment and are valid within the following ranges:

- v between 13.6 m/s and 28 m/s.
- w between $w_1 = 2.64$ m/s and $w_2 = 10.6$ m/s,
  corresponding to 13.3 rev/s and 53.2 rev/s,
  or 800 rev/min and 3250 rev/min, respectively.

The trajectory of the tennis ball is in three-dimensional space and is described by Newton's equation of motion with the three acting forces $\vec{G}, \vec{D}, \vec{M}$:

$$m \cdot \ddot{\vec{r}}(t) = \vec{G} - D\frac{\vec{v}}{v} + M\frac{\vec{\omega} \times \vec{v}}{\omega \cdot v}. \tag{5.1}$$

The initial conditions are

$$\vec{r}(0) = \vec{r}_0, \qquad \dot{\vec{r}}(t) = \vec{v}_0.$$

In the following, we consider the special case of plane trajectories in the (x, z)-plane, where $\vec{\omega}$ is perpendicular to the plane. Hence $\vec{\omega}$ is perpendicular to the initial velocity $\vec{v}_0$ and parallel to the y-axis.

From (5.1), after dividing by m and subsequently replacing $\alpha = \dfrac{\rho \pi d^2}{8m}$, $v = \sqrt{\dot{x}^2 + \dot{z}^2}$, and $\eta = \pm 1$ for either topspin or slice, we obtain

$$\begin{pmatrix} \ddot{x} \\ \ddot{z} \end{pmatrix} = \begin{pmatrix} 0 \\ -g \end{pmatrix} - \alpha \cdot C_D \cdot v \begin{pmatrix} \dot{x} \\ \dot{z} \end{pmatrix} + \eta \cdot \alpha \cdot C_M \cdot v \begin{pmatrix} \dot{z} \\ -\dot{x} \end{pmatrix}, \tag{5.2}$$

because

$$\left| v \cdot \begin{pmatrix} \dot{x} \\ \dot{z} \end{pmatrix} \right| = \left| v \cdot \begin{pmatrix} \dot{z} \\ -\dot{x} \end{pmatrix} \right| = v^2 .$$

Note that the Magnus force is perpendicular to the velocity vector.
For a topspin shot ($\eta = 1$), it points downwards (as shown in the previous figure), for a slice shot ($\eta = -1$), it points upwards.

Initial conditions:

$$x(0) = 0, \quad z(0) = h, \quad \dot{x}(0) = v_0 \cos\theta, \quad \dot{z}(0) = v_0 \sin\theta$$

with h = launch height, v_0 = initial speed, and θ = launch angle above the horizontal.

5.7.2 Data and Comparison of Forces

(a) Diameter of the ball $d = 6.3$ cm $= 0.063$ m, mass of the ball 0.050 kg,
   $\rho = 1.29$ kg/m³.
(b) The initial speed was always chosen to be $v_0 = 25$ m/s.
(c) For the circumferential speed the two values $w_1 = 2.64$ m/s and
   $w_2 \approx 4 \cdot w_1$ were chosen, corresponding to 800 rev/min and 3250 rev/min.

The following diagram shows the drag forces D computed for $w = w_1$ and $w = 4w_1$:

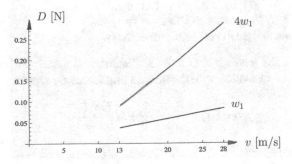

The following diagram shows the Magnus forces M computed for $w = w_1$ and $w = 4w_1$:

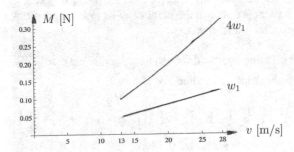

For comparison, the ball's weight is $G \approx 0.5$ N.

5.7.3 Mathematica Program for Computations and Diagrams

Let us rewrite (5.2) as a system of first order differential equations with the time-varying velocity components $v_x(t), v_z(t)$. The same notations will be used in the following program on the lines led by eq1, ..., eq4.

$$\left. \begin{aligned}
\dot{x}(t) &= v_x(t) \\
\dot{v}_x(t) &= -\alpha \cdot C_D \cdot v(t) \cdot v_x(t) + \eta \cdot \alpha \cdot C_M \cdot v(t) \cdot v_z(t) \\
\dot{z}(t) &= v_z(t) \\
\dot{v}_z(t) &= -g - \alpha C_D \cdot v(t) \cdot v_z(t) - \eta \cdot \alpha \cdot C_M \cdot v(t) \cdot v_x(t)
\end{aligned} \right\}.$$

The following program traject[w, θ] calculates the trajectory for arbitrary initial values and delivers the diagrams:

- the two quantities w, θ are input parameters.
- v_0 may be set on line 2 and the launch height z[0] on the line led by in2.

For slice trajectories, set $\eta = -1$ on line 2.
NDSolve[...] computes the solution, ParametricPlot[...] generates the diagram.

```
traject[w_, θ_] :=
(g = 9.81; d = 0.063; m = 0.05; ρ = 1.29; α = Pi * ρ * d²/(8m); η = 1; v0 = 25;
v[t_] = (x'[t]² + z'[t]²)∧(1/2);
Cd = 0.508 + (1/(22.503 + 4.196(v[t]/w)∧(5/2))∧(2/5);
Cm = 1/(2.202 + 0.981(v[t]/w));
eq1 = x'[t] == vx[t];
eq2 = vx'[t] == -Cdαv[t]vz[t];
eq3 = z'[t] == vz[t];
eq4 = vz'[t] == -g - Cdαv[t]vz[t] - ηCmαv[t]vx[t];
in1 = x[0] = 0;
in2 = z[0] = 1.0;
in3 = vx[0] = v0Cos[θ];
in4 = vz[0] = v0Sin[θ];
sol = Flatten[NDSolve[eq1, eq2, eq3, eq4, in1, in2, in3, in4,
{x[t], z[t], vx[t], vz[t]}, {t, 0, 4}]];
{xx[t_], zz[t_]} = {x[t], z[t]}/.sol;
{vvx[t_], vvz[t_]} = {vx[t], vz[t]}/.sol;
speed[t_] := (vvx[t]² + vvz[t]²)(1/2);
ParametricPlot[{xx[t], zz[t]]}, {t, 0, 1.3}, AspectRatio− > Automatic,
AxesOrigin− > {0, 0}, Ticks− > {{0, 5, 10, 18, 20, 23}, {1, 2}}])
```

The topspin trajectories shown below were calculated and plotted by the following commands:

```
T1 = traject[2.64, 17Degree];
T2 = traject[10.55, 17Degree];
Show[{T1, T2}, PlotRange− > All]
```

5.7.4 Topspin Trajectories

Launch angle = $17°$, launch height = 1 m, $\eta = 1$:

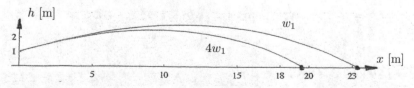

- for $w = w_1$: flight time $t = 1.30$ s, impact velocity $v = 15.3$ m/s.
- for $w = w_2 \approx 4w_1$: flight time $t = 1.10$ s, impact velocity $v = 14.8$ m/s.

5.7.5 Slice Trajectories

Launch angle = 0°, launch height = 1.4 m, $\eta = -1$:

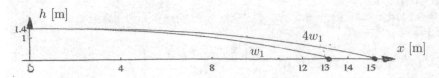

- for $w = w_1$: flight time $t = 0.615$ s, impact velocity $v = 19.0$ m/s.
- for $w = w_2 \approx 4w_1$: flight time $t = 0.76$ s, impact velocity $v = 16.6$ m/s.

5.7.6 Comparing Topspin Trajectory with Trajectory in a Vacuum

The parabolic trajectory in a vacuum

$$\begin{pmatrix} x(t) \\ y(t) \end{pmatrix} = \begin{pmatrix} \cos(17°) \cdot 25 \cdot t \\ 1 + \sin(17°) \cdot 25 \cdot t - \frac{g}{2}t^2 \end{pmatrix}$$

with the same launch height = 1.0 m, launch angle = 17°, and initial velocity $v_0 = 25$ m/s as in the case of topspin would result in a flight time of $t = 1.617$ s, an impact velocity of $v = 25.4$ m/s and a long flight distance of $x = 38.7$ m.

Remark: The total length of a tennis court is 23.8 m.

5.8 Mathematical Model for Skydiving

We wish to find a mathematical model for computing the velocity function $v(t)$ of a complete skydive from leaving the airplane to landing on the ground.

We assume that the descent is vertical and that the magnitude of the drag W varies with speed v as follows:

$$W(v) = \rho_L \cdot A \cdot C_W \cdot v^2 = k \cdot v^2.$$

Here

(a) $\rho_L = 1.29$ kg/m^3 is the air density,
(b) A is the silhouette area of the body, by which we mean the cross-sectional area orthogonal to the direction of airflow in square meters,
(c) C_W is the drag coefficient, a dimensionless number that depends on the shape.

Assumptions:

- The skydiver including equipment has a total mass of $m = 80$ kg.
- During free fall, when the body takes a horizontal position, the silhouette area is $A_1 = 0.72$ m^2, and $C_{W1} = 0.28$.
- After opening the parachute, the silhouette area is $A_2 = 17$ m^2 and $C_{W2} = 0.9$.

Newton's equation of motion for this one-dimensional problem is

$$m\dot{v} = mg - W(v) \quad \Longrightarrow \quad \dot{v} = g - \frac{1}{m}W(v) .$$

5.8.1 Analytical Model

The analytical model involves an autonomous differential equation distinguishing two cases:

- **Before opening the parachute**

$$\dot{v} = g - k_1 v^2 \text{ where } k_1 = \frac{\rho_L \cdot A_1 \cdot C_{W1}}{m} = \frac{1.29 \frac{\text{kg}}{\text{m}^3} \cdot 0.72 \text{ m}^2 \cdot 0.28}{80 \text{ kg}} = 0.003251 \text{ m}^{-1} .$$

For $g - k_1 v^2 = 0$, we obtain a constant particular solution

$$v_\infty = \sqrt{\frac{g}{k_1}} = 54.9 \text{ m/s} .$$

This is the limiting velocity, where drag and weight are equal.

An autonomous differential equation is separable. Therefore,

$$\int \frac{dv}{g - k_1 v^2} = \int 1 \cdot dt + C .$$

Because $v_\infty^2 = \frac{g}{k_1}$ and $v < v_\infty$, the integral is

$$\frac{1}{k_1} \int \frac{dv}{v_\infty^2 - v^2} = t + C \Longrightarrow \frac{1}{2k_1 v_\infty} \ln \frac{v_\infty + v}{v_\infty - v} = t + C,$$

obtained from an integration table or by the command NSolve[...].
Exponentiating

$$\ln \frac{v_\infty + v}{v_\infty - v} = \lambda_1 \cdot t + C \text{ yields} \quad \frac{v_\infty + v}{v_\infty - v} = \beta e^{\lambda_1 t} \quad \text{where } \lambda_1 = 2v_\infty \cdot k_1 = 0.3568.$$

If the skydiver drops from the airplane at time $t = 0$, then the initial condition is $v(0) = 0$, implying $\beta = 1$. Solving for v results in

$$v_\infty + v = (v_\infty - v)e^{\lambda_1 t} \Rightarrow v(1 + e^{\lambda_1 t}) = v_\infty e^{\lambda_1 t} - v_\infty \Rightarrow v(t) = \frac{e^{\lambda_1 t} - 1}{e^{\lambda_1 t} + 1} \cdot v_\infty .$$

Hence the solution is

$$v(t) = \frac{1 - e^{-\lambda_1 t}}{1 + e^{-\lambda_1 t}} \cdot v_\infty \quad \text{where } \lambda_1 = 0.3568 .$$

- **After opening the parachute**

$$\dot{v} = g - k_2 v^2 \text{ where } k_2 = \frac{\rho_L \cdot A_2 \cdot C_{W2}}{m} = \frac{1.29 \text{ kg/m}^3 \cdot 17 \text{ m}^2 \cdot 0.90}{80 \text{ kg}} = 0.2467 \text{ m}^{-1}.$$

For $g - k_2 v^2 = 0$, we obtain the constant particular solution

$$v_L = \sqrt{\frac{g}{k_2}} = 6.30 \text{ m/s} .$$

This is the landing velocity, where drag and weight are equal.

The computation is similar to that of the previous case of before opening the parachute. However, here we have a decelerating motion. The velocity v decreases from v_∞ to v_L. Hence $v > v_L$, leading to a different integral

$$\frac{1}{k_2} \int \frac{dv}{v_L^2 - v^2} = t + C \implies \frac{1}{2k_2 v_L} \ln \frac{v + v_L}{v - v_L} = t + C .$$

$$\ln \frac{v + v_L}{v - v_L} = \lambda_2 \cdot t + C \implies \frac{v + v_L}{v - v_L} = \gamma e^{\lambda_2 t} \text{ where } \lambda_2 = 2v_L \cdot k_2 = 3.12$$

Because $v(0) = v_\infty$, we have

$$\gamma = \frac{v_\infty + v_L}{v_\infty - v_L} .$$

Solving for v yields

$$v + v_L = (v - v_L)e^{\lambda_2 t}\gamma \implies v_L(1 + e^{\lambda_2 t}\gamma) = v(e^{\lambda_2 t}\gamma - 1) \implies v(t) = \frac{e^{\lambda_2 t}\gamma + 1}{e^{\lambda_2 t}\gamma - 1} \cdot v_L.$$

Or, introducing a time-shift of 15 s after which $v = v_\infty$ is practically attained,

$$v(t) = \frac{\gamma + e^{-\lambda_2(t-15)}}{\gamma - e^{-\lambda_2(t-15)}} \quad \text{where} \quad \lambda_2 = 3.12 \quad \text{and} \quad \gamma = \frac{v_\infty + v_L}{v_\infty - v_L}.$$

The following Mathematica code generates the function $v(t)$ with the two cases before and after opening the parachute at time $t = 15$ s, and subsequently plots its graph:

```
eig1[t_] := Exp[-0.3568t];
eig2[t_] := Exp[-3.120t];
vlimit = 54.9; vlanding = 6.3;
gamma = (vlimit + vlanding)/(vlimit - vlanding);
v[t_] := 54.9(1 - eig1[t])/(1 + eig1[t]) /; t <= 15;
v[t_] := 6.3(gamma + eig2[t - 15])/(gamma - eig2[t - 15]) /; t > 15;
Plot[v[t], {t, 0, 24}, AspectRatio- > 0.4, AxisOrigin- > {0, 0},
    PlotRange- > All, AxesLabel- > {"t  [s]", "v  [m/s]"}}];
```

Here is the graph of v:

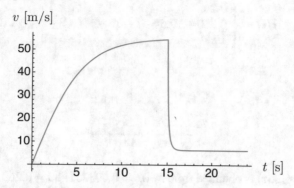

The velocity changes abruptly at the opening time $t = 15$ s. Zooming in for the time interval $[15.0001 \text{ s}, 15.2 \text{ s}]$ yields the following diagram:

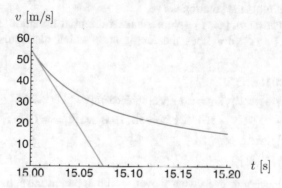

The slope of the tangent (orange) in the figure corresponds to a maximum acceleration of $\approx 53/0.08 > 600 \text{ m/s}^2$. This is more than $60\ g$, which no human being can survive. Hence our model fails during the opening process! The reason is that the drag doesn't change abruptly, as the parachute needs some time to deploy.
In the following numerical treatment, we shall improve the above analytical model.

5.8.2 Numerical Model

In the previous subsection we assumed that the parachute opens suddenly. This defect is now corrected by assuming that the opening process takes 3.0 s and that the drag during the time interval $[15.0 \text{ s}, 18.0 \text{ s}]$ changes continuously.
In this setting, the Initial value problem is $\quad m \cdot \dot{v} = m \cdot g - k(t) \cdot v^2 \quad$ with $v(0) = 0$, where $k(t)$ is given by the continuous, piecewise linear function

$$k(t) = \begin{cases} 0.26 & \text{if } t < 15 \\ 0.260 + 19.54 \cdot (t-15)/3 & \text{if } 15 \le t \le 18 \\ 19.8 & \text{if } t > 18 \end{cases}.$$

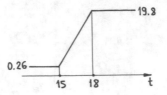

The top line of $k(t)$ is obtained from $k_1 \cdot m = 0.003251 \cdot 80 = 0.260$, the bottom line of $k(t)$ is obtained from $k_2 \cdot m = 0.2467 \cdot 80 = 19.8$.

The following Mathematica code numerically solves the initial value problem with little effort and plots the solution curve.
Note that in NDSolve[...] (Numerical Differential Equation Solver) the command Which[...] describes the factor $k(t)$, which distinguishes the various cases.

```
m = 80; g = 9.81;
sol = NDSolve[{mv1'[t] == mg − v1[t]² Which[t < 15, 0.26,
t < 18, 0.26 + 19.54(t − 15)/3, t >= 18, 19.8], v1[0] == 0}, v1, {t, 0, 22}];
v1[t_] = v1[t]/.sol;
Plot[v1[t], {t, 0, 22}];
```

The solution is smooth everywhere, even at the apparent cusp, as is illustrated on the right-hand diagram by zooming in towards $14.95 < t < 15.05$.

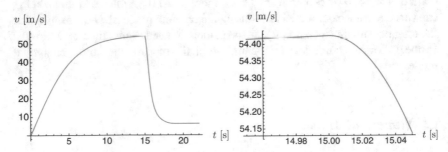

Finally, we wish to compute and plot the acceleration function

$$a(t) = \dot{v}(t) = g - v^2.$$

```
acc[t_] := g − v1[t]²
Which[t < 15, 0.26,  t < 18, 0.26 + (19.8 − 0.26)(t − 15)/3,  t >= 18, 19.8]/m;
Plot[acc[t], {t, 0, 22}], AxesLabel− > {"t  [s]", "v  [m/s]"}
```

Here is the graph of the acceleration a:

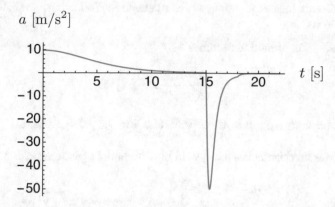

Its maximum value is approximately $5g$. Within about 0.4 s the skydiver experiences an acceleration increase from 0 to $5g$.

Of course, the question remains whether the opening process could be modeled better still. In order to adjust the function $k(t)$, more detailed investigation on the opening process would be required.

The numerical solution shows rather nicely that it is decidedly better than the analytical solution. Besides, it is more flexible in terms of modifications. Last, and most of all, it is obtained with significantly less effort!

5.9 Near-Earth Satellite Orbits

Due to its rotation, the earth is slightly flattened at the poles and bulges at the equator. The distance of the poles to the earth's center is about 20 km less than the equatorial radius R = 6378 km. The difference is just about 0.31%.

Two bodies with masses M and m attract each other. The magnitude of the force of attraction is

$$|\vec{F}| = G\frac{M \cdot m}{r^2} \quad \text{where } G = 6.6741 \cdot 10^{-11}\frac{\text{m}^3}{\text{kg} \cdot \text{s}^2}.$$

Assuming that the origin of the reference frame is at the center of the earth, the earth's mass is $M = 5.9722 \cdot 10^{24}$ kg, and a satellite of mass m is at a variable position $\vec{r}$, then Newton's equation of motion says

$$m\ddot{\vec{r}}(t) = -\vec{F} \quad \Longrightarrow \quad \ddot{\vec{r}}(t) = \frac{-G \cdot M}{r^3} \cdot \vec{r},$$

whose Kepler solutions we have discussed earlier.

The earth's bulge for near-earth satellite orbits gives rise to a perturbation force.[7]

---

[7] Approximation by spherical functions.

Because the earth bulges symmetrically with respect to the equatorial plane, the satellite orbits are located in the equatorial plane. In the following, we shall examine only such orbits.

Newton's equation of motion becomes

$$\ddot{\vec{r}}(t) = \left(\frac{-\mu}{r^3} - J\frac{\mu R^2}{r^5} \right) \cdot \vec{r}$$

where the constants are $\mu = M \cdot G = 398600 \text{ km}^3/\text{s}^2$, $J = 1.083 \cdot 10^{-3}$.

Thus in Cartesian coordinates the system of differential equations reads

$$\left. \begin{aligned} \ddot{x}(t) &= -\mu \frac{x}{r^3} - \mu J R^2 \frac{x}{r^5} \\ \ddot{y}(t) &= -\mu \frac{y}{r^3} - \mu J R^2 \frac{y}{r^5} \end{aligned} \right\} \quad \text{where } r = \sqrt{x^2 + y^2}.$$

Now an excerpt from a project:[8]

The assignment was to compute and graphically represent a near-earth satellite orbit in the equatorial plane taking into account the earth's bulge and to subsequently determine how much the perihelion precession per revolution is.

We begin with calculating the orbit. Later we describe the diagram depicting the perihelion precession as a rosette-like orbit that is no longer closed. The individual loops are slightly disturbed ellipses, while perturbation is caused by the earth's bulge.

Constants:
```
Abtastwerte=5000, μ=398600, R=6378, J2=0,01623
```

Remark: To better represent the effect, $J2 = 0.01623$ was used, 15 times the correct value of $J1 = 0.001083$.

Coding the IVP:

```
Diff[vx0_, vy0_, x0_, y0_] =
{
```

$$x''[t] == -\mu * \frac{x[t]}{\left(\sqrt{x[t]^2 + y[t]^2}\right)^3} - \mu * J2 * R^2 * \frac{x[t]}{\left(\sqrt{x[t]^2 + y[t]^2}\right)^5},$$

$$y''[t] == -\mu * \frac{y[t]}{\left(\sqrt{x[t]^2 + y[t]^2}\right)^3} - \mu * J2 * R^2 * \frac{y[t]}{\left(\sqrt{x[t]^2 + y[t]^2}\right)^5},$$

$$x'[0] == vx0, \quad y'[0] == vy0, \quad x[0] == x0, \quad y[0] == y0$$

```
}
```

---

[8] My former students Reber and Dellenbach carried out a mathematics workshop project in the Department of Technology and Computer Science at the Bern University of Applied Sciences.

The solution with initial position $(x0, y0) = (6600\,\text{km}, 0\,\text{km})$ and initial velocity $(vx0, vy0) = (0\,\text{km/s}, 7\,\text{km/s})$ is calculated by

```
sol = Flatten[NDSolve[{Diff[6600,0,0,7]},{x[t],y[t]},{t,0,5000}]];
```

The semicolon at the end suppresses the output. The command for the following diagram (without the five marks for the perihelion points) is

```
orbit=ParametricPlot[Evaluate[{x[t], y[t]} /.sol],
{t,0,5000},AspectRatio->Automatic];
```

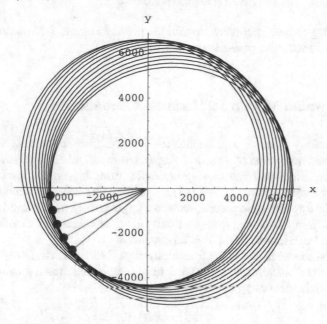

Procedure for finding the perihelion points:

(a) Extract 5001 discrete points $(x[t], y[t])$ for $t = 0, 1, 2, \ldots, 5000$ from the solution:
```
DiffMatrix=Table[NDSolve[{x[t],y[t]}/sol, {t,0,5000}]];
```

(b) Compute the 5001 distances $r(t) = \sqrt{x[t]^2 + y[t]^2}$:
```
NormMatrix=Table[Norm[DiffMatrix[[i]], {i,1,5000}]];
```

(c) Extract from these the positions of the five smallest distances:
```
MinimalPositionsMatrix=Ordering[NormMatrix,5]
```

(d) Finally access the five coordinate pairs and compute the intermediate angles by use of the inner product.

Comment on the diagram:
The rosette-shaped orbit with ellipse-like revolutions is not a closed curve.
The perihelion precession is $8.8°$ per revolution.

Flaw: The orbit is not always outside the earth. The initial conditions need to be changed suitably.

Remark: An investigation in [10] using the so-called *Averaging Method*[9] by Bogolyubov (1909–1992, Russian theoretical physicist) and Mitropolsky (1917–2008, Russian mathematician) concludes that a satellite orbit in the equatorial plane has a perihelion precession of $\Delta\varphi \approx 3\pi J (R/p)^2$ radians per revolution. In this context $p = \frac{b^2}{a}$ is the ellipse parameter with semiaxes $a \geq b$, and R is the earth's radius.

For a near-earth satellite orbit with $R \approx a \approx b$ (almost circular), we have

$$\Delta\varphi \approx 3\pi J \approx 1.02 \cdot 10^{-2},$$

corresponding to $3.67°$ per revolution. In the previous figure, as mentioned before, there are five successive perihelion point marks.

5.10 Brownian Motion and Langevin Equation

As a preparatory example to introduce the Kalman filter in the next section, we study the random motion of a particle suspended in a fluid. This erratic motion is caused by collisions with fast moving molecules from the surroundings. The Scottish botanist Robert Brown (1773–1858) discovered that pollen suspended in water, when viewed through a microscope, showed a jiggling motion, now known as Brownian motion. In the following we show that by use of statistics even a random motion like this can be described by differential equations.

The simplest model is due to the french physicist Paul Langevin(1872-1946): We consider the rectilinear motion of a particle of mass m in a liquid medium moving along the x-axis. According to Newton's law

$$m\ddot{x}(t) = F(t) - \rho \cdot \dot{x}(t). \tag{5.3}$$

Here F is a random force with mean zero and constant variance. In other words, F is a white noise. The last term expresses the assumption that the frictional force caused by the medium is proportional to the particle's speed. The parameter ρ depends on the viscosity of the liquid. Equation (5.3) is also called the Langevin Equation.

In order to solve it, we first multiply both sides by $x(t)$ to obtain

$$m\ddot{x}(t)x(t) = x(t)F(t) - \rho x(t)\dot{x}(t)$$

and rewrite it as
$$\frac{1}{2}m[(x^2)\dot{}]\dot{} - m\dot{x}^2 = xF - \frac{\rho}{2}(x^2)\dot{} , \tag{5.4}$$

because
$$m(x\dot{x})\dot{} - m\dot{x}^2 = (m\dot{x}^2 + mx\ddot{x}) - m\dot{x}^2 = m\ddot{x}x.$$

We take the mean values of (5.4) and use the fact that the mean value of x^2 is the variance $V(x)$ of x (if its mean is zero). Thus

---

[9] The Averaging Method is used in analytic perturbation theory for nearly periodic processes and is discussed in [3].

$$\frac{m}{2}\frac{d^2V}{dt^2} = 2\overline{E_{kin}} - \frac{\rho}{2}\frac{dV}{dt}.$$

We used that F is a white noise, so the mean of xF is zero. The term $\overline{E_{kin}} = \overline{\frac{m\dot{x}^2}{2}}$ is the mean kinetic energy. In statistical mechanics, the Equipartition Theorem relates it with the absolute temperature T of the system:

$$\overline{E_{kin}} = \frac{k_B}{2}T$$

where k_B is the Boltzmann constant. Therefore,

$$\frac{m}{2}\frac{d^2V}{dt^2} = k_BT - \frac{\rho}{2}\frac{dV}{dt}.$$

To solve this differential equation, we introduce $f(t) = \frac{dV}{dt}$ to obtain a linear autonomous first order differential equation:

$$\dot{f} = \frac{2k_BT}{m} - \frac{\rho}{m}f.$$

A particular solution is $f_\infty = \frac{2k_BT}{\rho}$, and the general solution is

$$f(t) = f_\infty + C \cdot e^{-\frac{\rho}{m}t}.$$

Therefore, the solution $V(t)$, satisfying $V(0) = 0$, for large t is approximated by

$$V(t) \approx t \cdot f_\infty = 2D \cdot t.$$

Accordingly, the mean square distance increases linearly with time t. This result was found by Albert Einstein, Marian Smoluchowski, and Paul Langevin around 1906. The diffusion coefficient D is given by the Stokes-Einstein Equation:

$$D = \frac{k_BT}{\rho} = \frac{k_BT}{6\pi\eta r}$$

where r is the hydrodynamic radius of a spherical particle and η is the viscosity of the medium.

5.11 Kalman Filter

The Kalman[10] filter estimates observables (e.g. signals) from indirect, inaccurate or uncertain measurements in the presence of white noise.

5.11.1 Theory

Let us start with the linear differential equation

$$\frac{dy(t)}{dt} = -ay(t) + w(t) \tag{5.5}$$

that describes a system where $a > 0$ is a parameter, $y(t)$ is the state of the system at time $t \in [0, \infty)$ or $t \in [0, t_e)$, $y(0) = y_0$ is the initial condition, and the excitation $w(t)$ is a white random process with mean zero and constant excitation variance Q.

Applications:

- When processing signals, such as speech, sound, and images, Equation (5.5) is a model for slowly changing random signals $y(t)$ that are assumed to be generated by a hidden random $w(t)$.
- The Langevin Equation (5.3) describing the Brownian motion has the form (5.5) with $y = \dot{x}$.
- In automotive engineering, (5.5) describes the motion of a damped car suspension system subject to random shocks $w(t)$ from the uneven road surface.
- Kalman filtering is also used in space travel for tracking space shuttles.

In view of a system given by (5.5), we wish to measure the state $y(t)$ as accurately as possible. However, every practical measurement $m(t)$ is affected by some white noise $n(t)$ with mean zero and constant measurement variance R due to imperfections of the measurement equipment. It is modeled by

$$m(t) = y(t) + n(t). \tag{5.6}$$

The Kalman filter solves the following problem: Given the system described in (5.5) and the measurements modeled by (5.6), find the best estimate $Y(t)$ of $y(t)$. It is clear that some kind of averaging over the values of $m(t)$ is required in order to filter out the noise $n(t)$. A good estimate is in the sense of least squares. In other words, the estimate $Y(t)$ should have a minimum mean square error from the true value $y(t)$. The Kalman filter algorithm postulates the following linear differential equation for $Y(t)$:

$$\frac{dY(t)}{dt} = -aY(t) + k(t) \cdot [m(t) - Y(t)] \tag{5.7}$$

---

[10] Rudolf E. Kálmán (1930–2016) was a Hungarian-born American mathematician. Educated at MIT, he was a professor at Stanford University and from 1973 on at ETH Zürich until his retirement in 1997. In 2009 he was awarded the National Medal of Science by then President Barack Obama.

where $k(t)$ is a yet unknown function of t. What is the motivation behind the differential equation (5.7)? It seems plausible that if no measurement is involved, in other words, if $k(t) = 0$, then the estimate $Y(t)$ should evolve in the same way as the mean of the process $y(t)$. This explains the first term on the right-hand side. The second term states that the difference $m(t) - Y(t)$ between the measurement $m(t)$ and the estimate $Y(t)$ is used to adjust the differential equation for $Y(t)$ by a proportionality factor $k(t)$. Let us denote the estimation error by

$$e(t) = y(t) - Y(t). \tag{5.8}$$

Subtracting (5.7) from (5.5), and noting that (5.6) and (5.8) hold, we obtain

$$\dot{y} - \dot{Y} = -a(y - Y) + w - k(m - Y) = -ae + w - k(y + n - Y) = -ae + w - k(e + n),$$

or, with $\dot{e} = \dot{y} - \dot{Y}$,

$$\frac{de(t)}{dt} = -[a + k(t)] \cdot e(t) + w(t) - k(t)n(t). \tag{5.9}$$

This describes the evolution of the estimation error $e(t)$. It can be shown by statistical considerations that the random term $w(t) - k(t)n(t)$ has variance $Q + k(t)^2 R$. It can also be shown that the mean square error $P(t)$ of $e(t)$, which is the same as the variance of $e(t)$, satisfies the differential equation

$$\frac{dP(t)}{dt} = -2[a + k(t)] \cdot P(t) + Q + k(t)^2 R \tag{5.10}$$

for any arbitrary function $k(t)$. In order to minimize the mean square error $P(t)$, the right-hand side of (5.10) must be minimized. This means that the quadratic expression in $k(t)$,

$$-2k(t)P(t) + k(t)^2 R$$

has to be minimized. Differentiation with respect to k (while holding t fixed) leads to $-2P(t) + 2k(t)R = 0$. Therefore,

$$k(t) = \frac{1}{R}P(t). \tag{5.11}$$

Substituting (5.11) into (5.10) results in

$$\frac{dP(t)}{dt} = -2aP(t) - \frac{1}{R}P(t)^2 + Q. \tag{5.12}$$

This nonlinear differential equation for the least mean square error $P(t)$ is of Riccati type.[11]

---

[11] Jacopo Francesco Riccati (1676–1754), a Venetian nobleman, declined various academic offers of universities in Italy and Russia, in order to devote himself to the study of mathematics at his home. He studied extensively the differential equation that bears his name. However, it was Euler who in 1760 found its solution.

Let us for a moment look at the **general Riccati equation**. It has the form

$$\frac{dy}{dt} = A(t)y^2 + B(t)y + C(t).$$

If a particular solution $y_1(t)$ is known, then substituting

$$y = y_1 + \frac{1}{u(t)}$$

into the Riccati equation yields

$$\dot{y} = \dot{y}_1 - \frac{\dot{u}}{u^2} = A\left(y_1 + \frac{1}{u}\right)^2 + B\left(y_1 + \frac{1}{u}\right) + C \implies -\frac{\dot{u}}{u^2} = 2Ay_1\frac{1}{u} + A\frac{1}{u^2} + B\frac{1}{u}$$

implying

$$\dot{u} = -[2y_1 A(t) + B(t)]u - A(t).$$

This is a linear differential equation in u. The general solution u produces the general solution y.

With this at hand, we are able to solve Equation (5.12). As was said, it is of Riccati type with the constant coefficients

$$A = -\frac{1}{R}, \quad B = -2a, \quad C = Q.$$

So, a particular solution of (5.12) is the constant positive solution

$$P_1 = -aR + R\sqrt{a^2 + \frac{Q}{R}} = (\lambda - a)R \quad \text{with} \quad \lambda = \sqrt{a^2 + \frac{Q}{R}}.$$

In order to find the general solution of (5.12) we follow the above Riccati procedure:

$$P(t) = P_1 + \frac{1}{u(t)},$$

$$\dot{u} = \left(\frac{2P_1}{R} + 2a\right)u + \frac{1}{R} = 2\lambda u + \frac{1}{R}.$$

The latter has the particular solution $u_p(t) = -\frac{1}{2\lambda R}$, which is constant. Hence the general solution is

$$u(t) = -\frac{1}{2\lambda R} + De^{2\lambda t}.$$

Finally, since $P_1 = (\lambda - a)R$, the general solution of Equation (5.12) is

$$P(t) = (\lambda - a)R + \frac{1}{-\frac{1}{2\lambda R} + De^{2\lambda t}}. \tag{5.13}$$

This entails

$$P_\infty = \lim_{t \to \infty} P(t) = (\lambda - a)R \text{ for } D \neq 0.$$

The integration constant D can be obtained from the initial condition $P(0) = P_0$. With the resulting function (5.13) in mind, it remains to solve (5.7) by use of (5.11):

$$\frac{dY(t)}{dt} = -aY(t) + \frac{1}{R}P(t)[m(t) - Y(t)]. \tag{5.14}$$

The solution, satisfying the initial condition $Y(0) =$ mean of $y(0)$, is the desired optimal estimate $Y(t)$ of $y(t)$. Given the measurements $m(t)$ in numerical form, (5.14) is solved numerically.

The Kalman filter is widely used in engineering, finance, and statistics. Its main equations are reproduced here as a summary:

System:

$$\frac{dy(t)}{dt} = -ay(t) + w(t) \tag{5.15}$$

Measurements:

$$m(t) = y(t) + n(t) \tag{5.16}$$

Kalman filter:

$$\frac{dP(t)}{dt} = -2aP(t) - \frac{1}{R}P(t)^2 + Q \tag{5.17}$$

$$\frac{dY(t)}{dt} = -aY(t) + \frac{1}{R}P(t)[m(t) - Y(t)]. \tag{5.18}$$

In order to obtain the estimate $Y(t)$ of $y(t)$, first solve (5.17), then (5.18).

Here is a block diagram depicting the system with the Kalman filter:

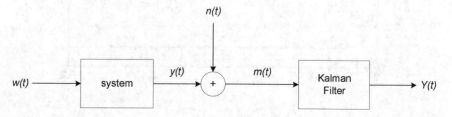

Remarks:

- The discrete-time version of the filtering algorithm was first derived by Kalman in 1960. The continuous-time filtering algorithm, Equations (5.17) and (5.18), was later given by Kalman and Bucy in 1961.
- Most implementations of Kalman filters are on digital computers using sampled measurements and discrete-time versions of the algorithm. They can be interpreted as recursive computations of the Gaussian linear least squares regression with sequential data.
- The Kalman filter algorithms remain valid also for systems with time-varying system parameter $a(t)$ and variances $Q(t)$ and $R(t)$. In this case (5.17) is typically solved by numerical methods.

5.11.2 Simulation of an Example

Here we give a numerical example that will be solved by Euler's method with a time step size of $h = 10^{-3}$ s, which corresponds to a sampling frequency of 1 kHz. The following numerical values were chosen:

- System parameter $a = 1/\tau$ with time constant $\tau = 0.2$ s, i.e. $a = 5 \text{ s}^{-1}$,
- Excitation variance $Q = 1$ of the input $w(t)$,
- Measurement noise variance $R = 10^{-4}$ of the noise $n(t)$.

First, the solution (5.13) of (5.12) with initial value $P_0 = 0.1$ is evaluated for $0 \text{ s} \le t < 0.2$ s:

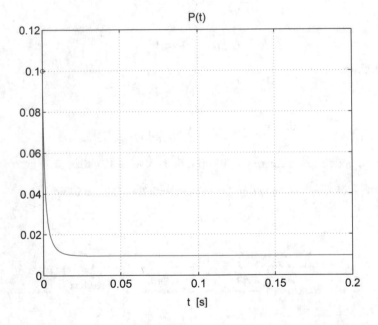

The fast convergence to $P_\infty = (\lambda - a)R = (\sqrt{5^2 + 10^4} - 5)10^{-4} \approx 0.0095$ holds for any initial value, as is evident from (5.13).

The following figure was generated by MATLAB:

First the simulation of a true signal $y(t)$ in black was generated by applying Euler's method to Equation (5.5) where $a = 5$ s^{-1}, $y(0) = y_0 = 0$, $w(t)$ is a random excitation signal with mean 0, and $Q/h = 1000$ is the excitation variance[12].

Then a random signal $n(t)$ was generated with mean $= 0$ and measurement variance $R/h = 0.1$ and added to $y(t)$ to obtain the noisy measurement $m(t)$ in blue.

Finally, the estimate $Y(t)$ in red for $y(t)$ was computed as the solution of Equation (5.14) using Euler's method with $Y(0) = \overline{y_0} = 0$ and $P_0 = 0.1$.

Note that the difference $n(t)$ between the blue and the black curves has a standard variation of $\sqrt{R/h} \approx 0.32$.

Also note that the difference $e(t)$ between the red and the black curves has a standard deviation of approximately $\sqrt{P_\infty} \approx 0.0975$.

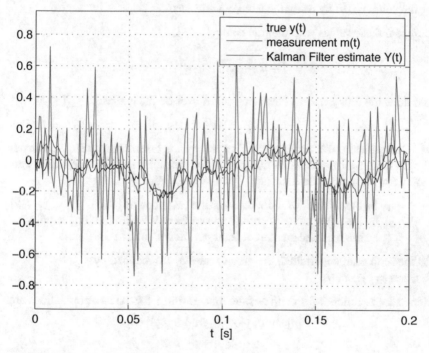

All three graphs consist of 200 discrete values, as $h = 10^{-3}$ s.

---

[12] The scaling factor $1/h$ is due to the discretization of the random signals $w(t)$ and $n(t)$.

5.12 Exercises Chapter 5

89. Euler's and Heun's methods. The IVP $y' = \sqrt{y}$ with $y(2) = 1$ is known to have the solution $y(x) = \frac{1}{4}x^2$.
Compute by use of $h = 0.1$ for $x = 2.10$ the y-values according to Euler's and Heun's methods. How do the two values compare with the exact value?

90. Model Problem. For the IVP $\quad y' = y$ with $y(0) = 1 \quad$ compute symbolically with one step of size h

(a) the value $y_e(h)$ by using Euler's method,
(b) the value $y_h(h)$ by using Heun's method,
(c) the value $y_{rk}(h)$ by using the Runge-Kutta method.

Compare these values with the true value

$$e^h = 1 + h + \frac{1}{2!}h^2 + \frac{1}{3!}h^3 + \frac{1}{4!}h^4 + \cdots$$

and draw the four graphs on the same set of axes on the interval $0 \le h \le 1$.

91. Competition of Two Species. The problem is taken from [28]. We consider a simple model for two species competing for the same food source, that means, a model for the interaction in an ecological system. For $i = 1, 2$, let

$\quad\quad p_i$ be the number of individuals of the two populations,
$\quad\quad a_i$ be the growth rate in percent, when there is unlimited food supply,
$\quad\quad b_i$ be the required amount of food per individual and time unit.

As the food source is limited, the growth rates a_i are reduced by
λ_i times $b_1 p_1 + b_2 p_2$.

Hence the equations of growth as functions of time t for the two populations are
$$\dot{p}_1 = [a_1 - \lambda_1(b_1 p_1 + b_2 p_2)] \cdot p_1$$
$$\dot{p}_2 = [a_2 - \lambda_2(b_1 p_1 + b_2 p_2)] \cdot p_2$$

This is a coupled autonomous nonlinear system of differential equations.
Simulate this population model for various sets of numerical parameters that satisfy the inequality $\dfrac{a_1}{a_2} > \dfrac{\lambda_1}{\lambda_2}$.

One set, for example, is

$$a_1 = 0.5 \quad \lambda_1 = 0.4 \quad b_1 = 0.005 \quad p_1(0) = 500$$
$$a_2 = 0.1 \quad \lambda_2 = 0.2 \quad b_2 = 0.001 \quad p_2(0) = 700$$

The two initital conditions $p_i(0)$ may be chosen arbitrarily.

Simulate different cases to show experimentally that, rather interestingly, the following is always true:

$$\lim_{t\to\infty} (p_1(t), p_2(t)) = \left(\frac{a_1}{\lambda_1 b_1}, 0\right)$$

and the right-hand side is the stable equilibrium point of the system of differential equations.

Thus one of the two populations becomes extinct ($p_2 = 0$). This is called the competitive exclusion principle, also referred to as the Volterra exclusion principle:
If one population grows faster than the other, then the latter will eventually die out. The theory on this can be found in [28] and in [2].

92. Model for Rocket Launch with Air Drag. We wish to set up a mathematical model for a vertical rocket launch up to a height of approximately 80 km in the form of a differential equation. Here the velocity $v(t)$ and the height $h(t)$ as functions of time are to be computed numerically for suitably chosen parameter values.

Typical data, such as void mass, fuel mass, thrust, C_W-value and the silhouette area can be gathered from the literature or the Internet. The air densitiy $\rho(h)$ is described in Exercise 93.

Hint: Compare with rocket launch without drag, Equation (3.4).

93. Model for Meteorite Orbits. Here are some data and pieces of vague information:
For computing the varying drag, the air density $\rho(h)$ as a function of the altitude h above sea level is required. Up to an altitude of about 100 km the exponential behavior

$$\rho(h) = \rho_0 e^{-h/H} \text{ where } H \approx 6000 \text{ m and } \rho_0 \approx 1.29 \text{ kg/m}^3$$

describes the standardized atmosphere quite well.
In the region of the earth's orbit meteorites have a maximum heliocentric velocity of approximately 42 km/s. As the orbital velocity of the earth is roughly 30 km/s, the velocities relative to the earth's atmosphere are between 12 km/s and 72 km/s.
The orbit depends on the entry velocity v_0 (assuming entry level $h_0 \approx 80$ km), the entry angle α, the weight G, and the geometrical shape of the meteorite, i.e. the silhouette area A and the drag coefficient C_W.

In the literature, impact velocities between 2 and 4 km/s are mentioned for large meteorites. Smaller meteorites experience strong braking forces, hence it is assumed that, even if falling freely, they may hit the earth's surface with impact velocities between 60 and 80 m/s.

The largest meteorite ever found, named the Hoba meteorite, is located in Namibia: The estimates about its weight range from 50 to 60 tons. It has the approximate shape of a cuboid (2.70 m × 2.70 m × 0.90 m), hit the earth some 80,000 years ago and still lies at the original location. Its age is estimated to be between 190 and 410 million years. It is composed of about 82% iron, 16% nickel, and 1% cobalt.

A fist-sized meteorite of diameter 10 cm and of similar composition as the Hoba meteorite has a mass of more or less 8 kg.

A meteorite may, on occasion, lose mass and size during its flight. This effect may be quite substantial. Smaller meteorites may completely burn up in the atmosphere. However, hardly any loss of material was found in the Hoba meteorite.

A **first test example** for the model is supplied by an excerpt from [31]:

"On the evening of April 6, 2002, numerous observers of Central Europe were offered an eerie spectacle: A fireball hissed across the sky, briefly illuminating it. In Bavaria the windows rattled, the floor trembled, the police department received dozens of calls. Depending on their mindsets, eyewitnesses reported falling space debris, a plane crash, or even a UFO.

But it soon turned out that the strange phenomenon was a meteorite. Surveillance cameras of the *European Fireball Network* had monitored a glowing trace of the celestial body. Based on the images, researchers were able to calculate the impact area. More than three months later, a fragment of 1.75 kg was found 6 km off Neuschwanstein Castle.

The meteorite, on the spot named after the castle, is a rarity: It is only the fourth of meteorites that were discovered by photos taken during their flights. Astronomers may use these images to reconstruct the entry path of the rock fragment — and hence its origin from the solar system.

The German Aerospace Center in cooperation with the Astronomical Institute of the Czech Academy of Sciences found that the glowing trace began at an altitude of 85 km, about 10 km northeast of Innsbruck. At an altitude of 21 km it flashed — apparently the rock of initially 300 kg broke into several fragments before its track disappeared west of Garmisch-Partenkirchen at an altitude of 16 km."

A **second test example** for the model:

On February 15, 2013, the **largest meteor in 100 years was observed in the Russian Ural region in the district of Chelyabinsk** with an entry angle of $16°$. It is well documented on the Internet, supported with photos and videos. It exploded and showered the region with many surviving meteorite fragments causing numerous damages.

Project outline:

(a) Establish an initial value problem that depends on the five above-mentioned parameters v_0, α, G, A, C_W.
(b) Solve numerically the IVP by aid of a Computer Algebra System that furnishes a time-dependent solution describing the path for appropriately chosen parameter values.

(c) Display a plot of the orbit and the time-dependent path. The range of the path from its initial altitude of 80 km to its impact on the ground can be read from the graph.
(d) Determine the time to impact.
(e) Compute the decreasing speed function $v(t)$, in particular the speed at impact.

Modeling also involves examining the C_W-values in order to choose them as realistically as possible.

Hint: Starting aids for modeling may be found in Sections 4.4, 5.7, and 5.8 on modeling with air drag, projectile trajectories with air drag, and skydiving.

94. Trajectory of a Tennis Ball. Establish a model for

(a) a tennis ball launched without spin, that is, without the Magnus force. Compare your result with those of Section 5.7. Convince yourself that the flight time is shorter than in the cases of slice or topspin balls.
(b) slice and topspin shots in the presence of tailwind or headwind.
(c) a slice shot, where the racket is hit almost parallel to the net with the effect that the ball swerves sideways, then vary the angle between the racket and the net.

In case (c) the trajectories are no longer planar.

95. Refined Predator-Prey Model. In the absence of predators, the prey population behaves according to the logistic growth model. Hence the equations (4.15) receive an additional term $-k \cdot h^2$ and become
$$\left. \begin{array}{l} \dot{h} = a \cdot h - k \cdot h^2 - b \cdot h \cdot f \\ \dot{f} = \qquad -c \cdot f \quad + d \cdot h \cdot f \end{array} \right\}.$$
Choose the parameters and initial conditions as realistically as possible, based on the data you acquired. Your choice should satisfy the condition $\frac{a}{k} \gg \frac{c}{d}$.

(a) Verify that $(h, f) = \left(\dfrac{c}{d}, \dfrac{a - kc/d}{b} \right)$ is a fixed point of the system.
(b) Generate various graphical solutions with various initial conditions by using the program code for (4.15) while including the additional term. As time $t \to \infty$, what do you find?

96. Riccati Differential Equation. Compute the general solution of

(a) the logistic equation $y' = ay - by^2$ where $a \gg b > 0$,
(b) the equation $y' = -y^2 + \frac{2}{t^2}$ by use of the particular solution $y_1 = \frac{2}{t}$.

Chapter 6
Climate Change and Epidemics

6.1 Climate Change

6.1.1 The Zero-Dimensional Energy Balance Model

Here we introduce the simplest model, also known as the **Zero-Dimensional Energy Balance Model**, where **solely globally averaged quantities** are used.

The theory originates from independent works of Budyko (1920–2001)[1] and Sellers (1955–2016).[2]

The heat equation of a body of any arbitrary shape is given by

$$\frac{\mathrm{d}Q}{\mathrm{d}t} = P_{\mathrm{gain}} - P_{\mathrm{loss}}.$$

The quantity of heat is $Q = C \cdot M \cdot T$ where T is the temperature in Kelvin, M the mass of the heat exchanging body, and C its mean specific heat capacity with dimension J/(kg K). Next we determine the power of heat gain P_{gain} and heat loss P_{loss} of the earth. The total power absorbed by the earth equals the short wave radiation received from the sun, hence

$$P_{\mathrm{gain}} = \pi R^2 \cdot S_0 \cdot (1 - \alpha).$$

Here πR^2 is the cross-sectional area of the earth, $S_0 = 1370$ W/m^2 is the total power per square meter received from the sun in the form of radiation (Total Solar Irradiance, TSI) and $\alpha \approx 0.32$ is the fraction that is reflected back into space, the so-called planetary albedo.

---

[1] Mikhail Ivanovich Budyko was a Russian climatologist, geophysicist and geographer. He is considered one of the leading climate researchers. Numerous models and predictions on global warming were incited by his research.

[2] Piers John Sellers was a British-American climate researcher and astronaut. After his three space shuttle flights from 2002 to 2010 he became a director of the Earth Science Division at Goddard Space Flight Center of NASA.

© The Author(s), under exclusive license to Springer Nature Switzerland AG 2021
A. Fässler, *Fast Track to Differential Equations*,
https://doi.org/10.1007/978-3-030-83450-0_6

The total loss of power due to the emitted long wave radiation is given by the **Stefan-Boltzmann law**[3]
$$P_{\text{loss}} = 4\pi R^2 \cdot \varepsilon \cdot \sigma \cdot g \cdot T^4.$$

Here $\varepsilon = 0.97$ is the emissivity, $\sigma = 5.67 \cdot 10^{-8}\,\frac{\text{W}}{\text{m}^2\text{K}^4}$ the Boltzmann constant, T the temperature on the earth's surface in Kelvin, and g the crucial radiation factor, which can be influenced and models the greenhouse gas effects.[4]

About $\frac{3}{4}$ of the earth's surface is covered with water that has a specific heat capacity of $4187\,\frac{\text{J}}{\text{kg}\cdot\text{K}}$. Considering the fact that some materials of the earth's crust have lower values, the global mean specific heat capacity is estimated at $C \approx 3440\,\frac{\text{J}}{\text{kg}\cdot\text{K}}$. Hence the temperature varies with time t, in seconds, according to the equation

$$4\pi R^2 \cdot \Delta R \cdot \rho \cdot C \cdot \frac{\mathrm{d}T}{\mathrm{d}t} = P_{\text{gain}} - P_{\text{loss}}.$$

Here ρ is the average density of the layer ΔR of the earth's surface that is involved in the heat exchange and is estimated at $\rho \approx 1500\ \text{kg/m}^3$.

Dividing by $4\pi R^2$ yields a nonlinear autonomous differential equation for the function $T(t)$ that is also found in [15], Subsection 3.2.1:[5]

$$C_E \cdot \frac{\mathrm{d}T}{\mathrm{d}t} = \frac{S_0}{4}(1 - \alpha) - \varepsilon \cdot \sigma \cdot g \cdot T^4.$$

The constant $C_E = \Delta R \cdot \rho \cdot C$, measured in J/(m^2 K), may be interpreted as the specific heat capacity per m^2.

According to the findings of the climate sciences the prevailing power of radiation $\frac{S_0}{4}(1 - \alpha)$ in the future will increase by the so-called **radiative forcing**

$$\Delta F = 5.35\,\frac{\text{W}}{\text{m}^2} \cdot \ln\left(\frac{c}{c_0}\right)$$

caused by feedbacks in the atmosphere due to greenhouse gases.

---

[3] Josef Stefan (1835–1893) was an Austrian mathematician and physicist with Slovenian roots, Ludwig Boltzmann (1844–1906) was an Austrian physicist.

[4] The scaling factor g is a result of the fact that the emission temperature in the real atmosphere, i.e. the temperature that is observed in the universe, does not correspond to the temperature on the earth's surface.

   The reason is that a considerable portion of the infrared radiation emitted by the earth's surface is trapped in the atmosphere, as it is absorbed by greenhouse gases, mainly steam and carbon dioxide CO_2, in the atmosphere. These particles radiate the absorbed energy in all directions, also back down towards the earth, thus heating up the earth's surface and, via convection, the entire atmosphere as well.

   In the convective atmosphere the temperature falls with rising altitude (see Section 3.11.2), mainly because the air expands adiabatically with decreasing pressure.

   Due to infrared radiation back into the universe, which depends on the altitude, the earth system can maintain its balance between incoming and outgoing radiation even in the presence of greenhouse gases. With the current concentration of greenhouse gases, infrared radiation into the universe takes place at an altitude of 5 km. In our simple model of energy balance we account for the heat loss by infrared emission by using the radiation factor $g < 1$. It should be noted that its exact value depends on many feedbacks of the earth system, for instance on the increase of the amount of steam in the atmosphere that heats up the earth.

[5] Reference provided by Prof. Dr. Stefan Brönnimann, Climatology Group of the Institute of Geography at the University of Bern.

Here $c_0 = 280$ ppm [6] is the relative pre-industrial reference pollution of the atmosphere by CO_2 particles and c is the current relative pollution.

Hence the autonomous differential equation with radiative forcing becomes

$$C_E \cdot \frac{dT}{dt} = \frac{S_0}{4}(1-\alpha) + \Delta F - \varepsilon \cdot \sigma \cdot g \cdot T^4. \tag{6.1}$$

The following **Keeling curve**[7] shows the c-values in the years 1958 to 2020 and is reproduced from [50]:

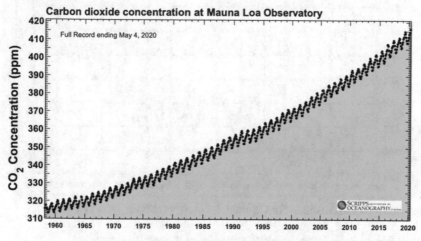

Monthly mean CO2 concentrations measured at Mauna Loa Observatory, Hawaii, from 1958 through to 4 May 2020.

During the short time period of 60 years the c-value increased by roughly 30 %!

The total concentration c of *all* greenhouse gases (carbon dioxid, methane,....) including cooling aerosols reached the value of $c = 457$ ppm in CO_2 equivalents in the year 2018 (see [42]).

In 2018, the Intergovernmental Panel on Climate Change IPCC (in German *Weltklimarat*) launched a special report [43] on the target of a global warming of $1.5°C$. For the calculations, we use the estimated mean value $c = 467$ ppm for the time span between the years 2018 and 2026. Therefore, the mean radiative forcing is

$$\Delta F = 5.35 \frac{W}{m^2} \cdot \ln\left(\frac{467}{280}\right) = 2.73 \frac{W}{m^2}.$$

Using the parameters given numerically in (6.1), we obtain the stationary solution as a function of the radiation factor g:

$$T_\infty(g) = \sqrt[4]{\frac{S_0(1-\alpha)/4 + \Delta F}{4 \cdot \varepsilon \cdot \sigma \cdot g}} = \sqrt[4]{\frac{1370 \cdot 0.68/4 + 2.73}{0.97 \cdot 5.67 \cdot 10^{-8}}} \cdot \frac{1}{\sqrt[4]{g}} = \frac{255.8}{\sqrt[4]{g}} \text{ K.} \tag{6.2}$$

---

[6] ppm is the abbreviation for parts per million, 1 ppm = 10^{-6}.

[7] Charles David Keeling (1928–2005), an American climate researcher, was a professor for chemistry at the Scripps Institution of Oceanography in San Diego, CA. He held guest professorships at the universities of Heidelberg (1969–1970) and Bern (1979–1980).

Let us analyze for various scenarios the crucial temperature rise Δ that is relevant to humans. According to the following diagram, the mean surface temperature $T_1 = (273.15 + 14.65)$ K $= 287.8$ K prevailed in the year 2020:

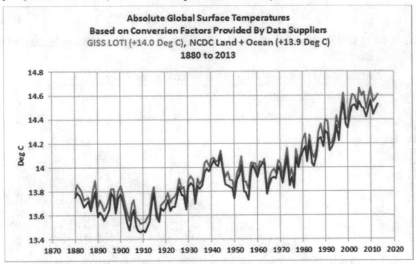

Absolute Global Surface Temperatures
Based on Conversion Factors Provided By Data Suppliers
GISS LOTI (+14.0 Deg C), NCDC Land + Ocean (+13.9 Deg C)
1880 to 2013

Source: European Institute for Climate & Energy (EIKE), Bob Tisdale.

Using Equation (6.2), we obtain the stationary temperature increase

$$\Delta_\infty(g) = \frac{255.8}{\sqrt[4]{g}} - 287.8 \quad \text{in } °C.$$

The equation $\Delta_\infty(g) = 0$ implies that for the g-value

$$g_0 = \left(\frac{255.8}{287.8}\right)^4 = 0.624$$

the temperature $T(t) = 14.65°C$ is constant in time.

Without greenhouse gases ($g = 1$), one would have
$\Delta_\infty(1) = 255.8 - 287.8 = -32° $ C !
Calculations for the two disputed temperature rises:

- $\Delta_\infty = 2°C$ implies $g_2 = 0.607$,
- $\Delta_\infty = 1.5°C$ implies $g_{1.5} = 0.611$.

Comparing $g_2 = 0.607 < g_{1.5} = 0.611 < g_0 = 0.624$, it becomes evident that a large g-value causes a small temperature increase Δ_∞.

In order to compute the function $T(t)$ quantitatively, the numerical value of C_E is required.
The penetration depth of light in water depends on its wave length. It varies from 5 m (for red light) to 60 m (for blue light). But the relevant water layer is mainly given by the convection depth that ranges from 10 m to over 100 m. It is called the mixed layer. Its average thickness is influenced by winds (impulse drive) and

changes in the water densities at the surface that are caused by warm and cold water flows (so-called buoyancy forces).

Taking into account that 1/4 of the earth's surface is covered with its crust, we estimate the thickness of the mean global mixed layer at $\Delta R \approx 40$ m and obtain

$$C_E \approx 40 \text{ m} \cdot 1500 \frac{\text{kg}}{\text{m}^3} \cdot 3440 \frac{\text{J}}{\text{kg} \cdot \text{K}} = 2.06 \cdot 10^8 \frac{\text{J}}{\text{m}^2 \cdot \text{K}}.$$

6.1.2 Linearized Model

Let $T(t) = T_1 + \Delta(t)$ where $T_1 = 287.8$ K. As Δ is small relative to T, we may use the binomial theorem and obtain a good estimate

$$T^4 = (T_1 + \Delta)^4 \approx T_1^4 + 4T_1^3 \cdot \Delta.$$

As $T(t)$ and $\Delta(t)$ only differ by the additive constant T_1, we have $\dfrac{d\Delta}{dt} = \dfrac{dT}{dt}$.
Hence (6.1) implies the linearized differential equation for $\Delta(t)$

$$C_E \cdot \frac{d\Delta}{dt} = A - B \cdot \Delta \qquad (6.3)$$

where

$$A(g) = \frac{S_0}{4}(1-\alpha) + \Delta F - \varepsilon \cdot \sigma \cdot T_1^4 \cdot g \quad \text{and} \quad B(g) = 4 \cdot \varepsilon \cdot \sigma \cdot T_1^3 \cdot g.$$

In order to compute the g-values, we shall substitute the numerical values for all parameters except g. Using

$$\frac{S_0}{4}(1-\alpha) + 2.73 = 235.6, \qquad \varepsilon \cdot \sigma = 0.97 \cdot 5.67 \cdot 10^{-8} = 5.500 \cdot 10^{-8}$$

$$T_1^4 = 6.8606 \cdot 10^9, \quad T_1^3 = 2.3838 \cdot 10^7$$

we obtain the linear relations

$$A(g) = 235.6 - 377.3 \cdot g \quad \text{and} \quad B(g) = 5.24 \cdot g.$$

Using $\dfrac{d\Delta}{dt} = 0$, the limit Δ_∞ of $\Delta(t)$ as $t \to \infty$, as a function of g is obtained as a zero of the right-hand side of (6.3):

$$\Delta_\infty(g) = \frac{A(g)}{B(g)} = \frac{235.6 - 377.3g}{5.24 \cdot g}, \quad \text{in } ^\circ\text{C}. \qquad (6.4)$$

Let us check Equation (6.4):
$\Delta_\infty = 0$. The equality $A(g) = 0$ is required. Once again, we obtain $g = g_0 = 0.624$. And $\Delta_\infty = 2^\circ\text{C}$ implies $g_2 = 0.6075$.

Calculating the temperature difference $\Delta(t)$ as a function of time for the scenario of global warming by 2.0 $^\circ$C from the year 2021 on:

Let us assume that the g-value is constant in time with $g = g_2$. Because of (6.3)

$$\frac{d\Delta}{dt} = \frac{A(g_2)}{C_E} - \frac{B(g_2)}{C_E} \cdot \Delta = \frac{6.39}{C_E} - \frac{3.19}{C_E}\Delta \qquad \text{where } \Delta(0) = 0.$$

Checking the target temperature: $6.39/3.19 = 2.00$.

The solution of the initial value problem is

$$\Delta(t) = 2.0 \cdot \left[1 - \exp\left(-\frac{3.19}{C_E} \cdot t\right)\right]. \tag{6.5}$$

Let us convert t seconds into τ years: 1 year $= 365 \cdot 24 \cdot 3600$ s $= 3.15 \cdot 10^7$ s.

The new constant in the exponent $-\dfrac{3.19 \cdot 3.15 \cdot 10^7}{2.06 \cdot 10^8} = -0.4878$ leads to the solution

$$\Delta(\tau) = \Delta_\infty(g) \cdot (1 - e^{-0.4878 \cdot \tau}), \tag{6.6}$$

where τ is measured in years and $\tau = 0$ corresponds to the year 2021.

The half-life is $T_{1/2} = 1.42$ years. After five half-lives $5T_{1/2}$, that is after 7.1 years, we have $\Delta = 1.94°$ C in the year 2028!

Remarks:

- With a time-dependent $g(t)$ that is possibly better adapted to real-life data and substituting the function $g(t)$ into (6.3), the problem can be modeled and numerically solved. Obviously also for cases with other targets, for instance, for 1.5° C.
- Within a time range of approximately 20 years, the computed constant C_E is reasonable. For larger time periods up to decades or centuries one must take into account that the warming of the seas takes place in much deeper layers, hence the quantity ΔR must be some multiple of 40 m.

6.1.3 Increase of Greenhouse Gases

While investigating the Keeling curve one might pose the question, whether or not the increase of the portion of carbon dioxide CO_2 in the atmosphere during the past decades was man-made. The answer unmistakeably is: Yes!

Reasoning:
Measurements of the concentration of the carbon C14 isotope relative to the concentration of C12 in the CO_2 atmosphere show that the ratio C14 / C12 in the past decades has dropped significantly.

The concentration in the atmosphere caused by cosmic radiation, however, was practically constant over thousands of years (radiocarbon dating is based on this fact, see Subsections 3.3.2 and 3.3.3).

Burning fossil fuels produces **no** C14, but solely C12. Intensive burning of fossil fuels in the past decades has remarkably diluted the concentration of C14 in the atmosphere.

This happened despite the fact that atomic bomb tests in the 1950s and 1960s have greatly increased the C14 / C12 ratio for some short periods of time!

Let us consider the following diagram, taken from [49]:

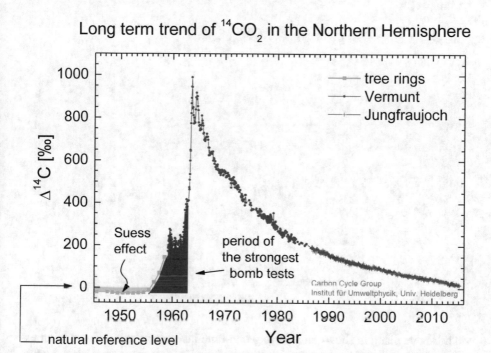

Here we define

$$\Delta^{14}C = f \cdot \frac{n^{14}}{n^C} - 1000. \tag{6.7}$$

$f = 8.19 \cdot 10^{14}$ is a dimensionless constant, n^{14} denotes the number of C14 atoms and n^C the number of CO_2 molecules in some unit volume of the atmosphere. Jungfraujoch, Switzerland is at an altitude of 3466 m above sea level, Vermunt, Austria at an altitude of 2000 m above sea level.

- Before 1950, the C14 concentration $r = \frac{n^{14}}{n^C} = 1.22 \cdot 10^{-12}$ was practically constant for thousands of years and $\Delta^{14}C = 0$ was valid.
- For double the C14 concentration $2r = 2.44 \cdot 10^{-12}$ we have $\Delta^{14}C = 1000$.
- For 1.5 times the C14 concentration $1.5r = 1.83 \cdot 10^{-12}$ we have $\Delta^{14}C = 500$.

It is remarkable that the known global data $\Delta^{14}C \approx 19\%$ and $\Delta^{14}C \approx 2\%$ in the years 1987 and 2017, respectively, practically coincide with the European values of the diagram.

Consequence: There is no doubt that the portion of carbon dioxide in the atmosphere is man-made! More information is found in [22].

By analyzing ice cores, the carbon dioxide concentration in the atmosphere can be traced back to hundreds of thousands of years. Here is one of several diagrams taken from [44]:

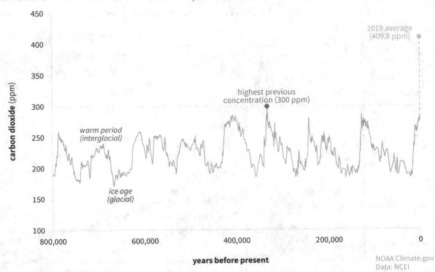

The above diagram shows an alarming one-time increase in recent times in an extremely short time interval of which the previous Keeling curve represents a zoom into the interval $(1950, 2016)$.

The argument that there have always been fluctuations is pointless, since these time intervals are of the order of 100,000 years, in contrast to a time interval that is only a few decades long!

Statement by Prof. Dr. Reto Knutti, ETH Zürich, member of the Intergovernmental Panel on Climate Change IPCC (Weltklimarat): "Climate change is developing into a climate crisis. Since the age of industrialization humans have increased the concentration of carbon dioxide in the air by 50 percent. Comparable values go back several millions of years, long before humans existed."[8]

In paleoclimatology, or the study of past climates, so called proxy data are in use to reconstruct past climate conditions. Examples for proxy data are:

[8] From the *Neue Zürcher Zeitung* NZZ, April 23, 2021.

(a) *Historical documents* contain a wealth of information about past climates.

(b) *Corals* build their hard skeletons from calcium carbonate, a mineral extracted from seawater. The carbonate contains isotopes of oxygen as well as trace materials that can be used to determine the temperature of the sea in which the coral grew.

(c) *Pollen*: All flowering plants produce pollen grains. The distinctive shape of pollen can be used to identify the type of plant from which they came. Since pollen grains are well preserved at the bottom of a lake or sea, an analysis of them in each layer tells us what kinds of plants were growing at the time the sediment was deposited. Based on the types of plants, inferences about the climate of that area can be made.

(d) *Ice Cores* have distinct layers containing dust, air bubbles or isotopes of oxygen, depending on the surrounding environment. The data can be used to interpret the past climate of the area in the course of time.

(e) *Tree Rings*: Climate conditions influence tree-ring widths, the density of the wood and isotopic composition. Such data reflect the variation of the climate over thousands of years.

(f) *Ocean and Lake Sediments*: Huge amounts of sediments on the floor of oceans and lakes provide information about the environment in past times. Drilled cores of sediments, containing tiny fossils and chemicals, allow an interpretation of the past climate.

By analyzing records taken from these and other proxy sources, one can extend the understanding of climate far beyond the instrumental record. The following diagram is taken from the *Science* Journal (see [36]):

Paleoclimate context for future climate scenarios

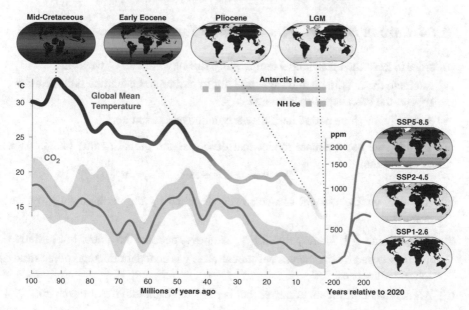

It shows the global mean temperature of the earth's surface and the CO_2 concentration way back to 100 million years. Blue bars indicate when there are well-developed ice sheets (solid lines) and intermittent ice sheets (dashed lines), according to previous syntheses. LGM stands for Last Glacier Maximum, SSP for Shared Socioeconomic Pathways, NH for Northern Hemisphere.

Past climate states were profoundly different from today. Their global mean temperatures, polar ice extents, regions of deep-water formation, vegetation types, patterns of precipitation and evaporation, and variability were all different. These differences are invaluable because they provide rich evidence of how climate processes operated across states that span the range of CO_2 concentrations associated with the future emission scenarios mentioned earlier.

On the right, curves of concentrations of CO_2 equivalents are shown for the three scenarios of so-called Shared Socioeconomic Pathways SSPs in the coming 200 years.
Under the sustainable SSP1-2.6 scenario, in which emissions are curtailed and become net negative by the end of the 21st century, CO_2 concentrations should stabilize near Pliocene levels. By contrast, under the fossil-fuel intensive SSP5-8.5 scenario, CO_2 concentrations approach or even exceed Eocene or mid-Cretaceous levels (see[27]).

The diagram shows indeed that the concentration of CO_2 equivalences in the atmosphere was below the value of about 476 ppm in the year 2021 during the last millions of years!
Remark: A comparable diagram in [45] contains more details than the diagram above.

6.1.4 Carbon Budget for the 2 Degree Target

In order to keep the mean global temperature rise to within 2° C, we shall
- calculate the maximum allowable amount of carbon to be emitted into the atmosphere, the so-called carbon budget.
- determine the time period until the carbon budget is used up.

To do this, we use a simple model that describes the global carbon budget as a function of time:
$$\Delta W = \Delta F - \Delta R \qquad \text{in } \frac{W}{m^2}.$$

(a) ΔW denotes the amount of net power absorbed by the earth (mainly by the oceans).

(b) $\Delta F = f \cdot \ln\left(\frac{c}{c_0}\right)$ where $f = 5.35\frac{W}{m^2}$, as above, denotes the amount of radiative forcing caused by the increase of greenhouse gas concentration and other atmospheric influences.

(c) ΔR denotes the amount of power that is reflected back into the atmosphere.

As a first approximation, ΔW and ΔR are proportional to the temperature increase ΔT:

$$\Delta W = \kappa \cdot \Delta T, \qquad \Delta R = \lambda \cdot \Delta T.$$

An analysis of the observed warming, taking into account the heat absorption of the oceans using climate models, provides the approximate parameters

$$\kappa = 0.6 \frac{W}{m^2 K} \text{ and } \lambda = 1.4 \frac{W}{m^2 K}.$$

Let us first compute the carbon dioxide concentration c_2 for the internationally agreed temperature increase $\Delta T = 2^\circ$ C:

$$\Delta F = \Delta W + \Delta R = f \cdot \ln \left(\frac{c}{c_0} \right) = (\kappa + \lambda) \cdot \Delta T.$$

However, the non-carbon dioxide radiative forcings[9] are not yet taken into account here. Assuming that according to the report [6] of the Intergovernmental Panel on Climate Change IPCC these cause about 0.5° C warming, we have to reduce ΔT by 0.5 for the following computation of $c = c_2$:

$$c_2 = c_0 \cdot \exp \left(\frac{(\kappa + \lambda) \cdot 1.5}{f} \right) = 280 \cdot \exp \left(\frac{(1.6 + 0.4) \cdot 1.5}{5.35} \right) = 491 \text{ ppm}.$$

It is known that increasing the value of c by 1 ppm, increases the pure carbon mass (without oxygen) M in the atmosphere by $a = 2.1$ Gt.[10]

To make it more precise, it should be noted here that the carbon dioxide (CO_2) mass is larger than the C mass by a factor of $(2 \cdot 16 + 12)/12 = 3.67$, because the relative atomic masses of oxygen O and carbon C are 16 and 12, respectively.

Carbon sinks in the ocean and forests remove an average of about half the emitted mass M_{em}. This means that only about half of the emissions remain in the atmosphere, a ratio that is known as the *airborne fraction* f_{air}. The value f_{air} remained fairly constant at 0.5 in the past few decades. However, model simulations show that with temperature increase ΔT, the airborne fraction f_{air} increases by about 5 % per degree of warming.

The relationship between the pure carbon mass M_{em} emitted in total and the carbon mass M remaining in the atmosphere is thus given by

$$\frac{M}{M_{em}} = f_{air} = 0.5 + 0.05 \cdot \Delta T.$$

For $\Delta T = 2^\circ$ C and $f_{air} = 0.6$, this means that 60 % of M_{em} remains in the atmosphere and 40 % is absorbed by the oceans and the biosphere.

---

[9] These are the contributions from methane CH_4 and nitrous oxide N_2O, i.e. greenhouse gases that act like CO_2, but enter the atmosphere through other sources and leave it in other ways.

[10] 1 Gt = 1 giga ton = 1 billion tons = 10^9 tons.

Hence, by increasing the concentration of the atmosphere from c_0 to c_2, the capacity of the total mass of emission is

$$M_{em} = \frac{a \cdot (c_2 - c_0)}{0.6} = \frac{2.1 \text{ Gt/ppm} \cdot (490.6 - 280) \text{ ppm}}{0.6} = 737 \text{ Gt}.$$

About 540 Gt were already emitted by the year 2020. Hence the budget that is allowed to be emitted thereafter is

$$\Delta M_{em} = (737 - 540) \text{ Gt} = 197 \text{ Gt}.$$

With the current emission mass of about 10 Gt per year, the remaining time period after 2020 would be about 20 years.

It should be borne in mind that the underlying parameters in the computations are dimensioned in such a way that the probabilities for an effective temperature increase below 2.0° C are about 50 %.

For a higher probability of approximately 66 %, one must use $\kappa + \lambda = 1.8 \frac{W}{m^2 K}$ for the calculations. The remaining time span would then be merely 10 years!

Coronavirus Pandemic 2020 and Global Warming [11]

The International Energy Agency IEA estimates that carbon dioxide emissions in 2020 will decrease by 2.6 billion tons compared to 2019. This corresponds to a **decrease of 8 %** with an emitted carbon dioxide mass of a total of 33 billion tons.

Nevertheless, the carbon dioxide concentration still increases: The emissions have been accumulating in the atmosphere since the age of industrialization. This explains an apparent paradox: Although we will emit less in 2020 than in the year before, the carbon dioxide concentration in the atmosphere will be higher than ever before.

To prevent the earth from warming up by more than 1.5° C with a probability of 66 %, global emissions have to **decrease by 7.6 % annually for the next 10 years**, estimates the United Nations Environment Program.

The book [24] provides a good overview of the topic of climate change.

---

[11] Taken from the article *The virus does not save the climate* by R. Fultener, Neue Zürcher Zeitung NZZ of May 7, 2020.

6.2 Epidemiology

Throughout history, humankind has been confronted with epidemics and pandemics time and again. The Spanish flu killed 25 to 50 million people in 1918–1920. Most recently, SARS (2002, causing atypical pneumonia), the avian flu (1983 and 1997) and the swine flu (2009) emerged in Asia. Africa was hit by Ebola several times in 1976, 2014, 2016, 2018, 2020.
Humankind has been confronted with the devastating coronavirus causing COVID-19 since 2019/2020.

6.2.1 The SIR Model

The SIR model was developed by two Scottish epidemiologists W. O. Kermack and A. G. McKendrick in the year 1927. The extended SEIR model used for the coronavirus pandemic will be discussed later.
We distinguish three disjoint groups of people S, I, R from a time-constant population of N individuals:

- Group S consists of people who can be infected and who have not yet come into contact with the disease in question (*susceptible individuals*).

- Group I consists of the people infected with the disease (*infectious individuals*) who are contagious for the group S.

- Group R consists of the people who have become immune to the disease or who have died because of the disease (*removed individuals*).

The functions $S(t), I(t), R(t)$ denote the number of people in the respective groups S, I, R as time t varies.

The SIR model is governed by a system of non-linear differential equations

$$\dot{S} = -a \cdot \frac{I}{N} \cdot S \tag{6.8}$$

$$\dot{I} = a \cdot \frac{I}{N} \cdot S - b \cdot I \tag{6.9}$$

$$\dot{R} = b \cdot I \tag{6.10}$$

with the initial conditions $S(0) = S_0$, $I(0) = I_0$, $R(0) = R_0$. At the beginning of an epidemic, typically $S_0 \approx N$, $I_0 \approx 0$, $R_0 = 0$, which may no longer be the case in a second wave.

Here **a is the infection rate** and **b is the immunity rate**. These are mean values that are characterized by the virus.

The number of births and deaths is relatively small compared to the total population, hence neglected in epidemics that take place over a period of months.

The following diagram serves for a better understanding:

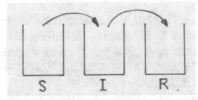

Individuals can only move from S to I and from I to R. In this model we assume that once an individual is immune then, at least for the duration of the epidemic, he or she remains immune.

Explaining the differential equations:

The probability that some person of group S (let's call her Eva) meets some other person of group I is I/N.

If Eva meets f people per day, the probability that she will meet an infected person per day is $f \cdot \frac{I}{N}$.

Assuming that an infection takes place with a transmission probability of p %, the probability that Eva will be infected on a certain day is $P = p \cdot f \cdot \frac{I}{N}$.

Finally, considering all S people, there are a total of $P \cdot S = p \cdot f \cdot \frac{I}{N} \cdot S = a \cdot \frac{I}{N} \cdot S$ people per day that are infected.

Provided with the factor -1, it is the negative term in the first equation (6.8) (decrease of S per day) and the positive term in the second equation (6.9) (increase of I per day) with the

$$\text{infection rate} \quad a = f \cdot p, \tag{6.11}$$

where f is the number of contacts per person per day and p is the probability of transmission.

The immunity rate b corresponds to the average probability per day that a person will become immune or die and can be viewed as the reciprocal of the duration T_i of infection or healing measured in days:

$$b = \frac{1}{T_i}. \tag{6.12}$$

Here healing means that the disease can no longer be passed on.

The term $b \cdot I$ equals the number of people per day who have become immune or have died. In the second equation (6.9), this term is subtracted as it decreases the rate of infected individuals. In the third equation (6.10), it appears as an increasing rate.

The right-hand sides in the system of differential equations add up to 0. Thus

$$S(t) + I(t) + R(t) = \text{constant} = S_0 + I_0 + R_0 = N.$$

The sum of the three functions equals the constant total population N.

The so-called **basic reproduction number** [12] is defined as follows:

$$R_0 = \frac{a}{b}.$$

An epidemic occurs if and only if

$$\dot{I}(0) = I_0 \left(\frac{a}{N} \cdot S_0 - b \right) > 0 \quad \Leftrightarrow \quad S_0 > \frac{N}{R_0}.$$

The initial value S_0 must therefore be greater than the so-called **threshold value** $\frac{N}{R_0}$.
Hence $R_0 > 1$ is necessary.

The infection rate a in the absence of immunized people is the probability that an infected person will infect another person on a given day. Therefore, because of

$$R_0 = a \cdot \frac{1}{b} = a \cdot T_i,$$

we have

$R_0 =$ **average number of people infected by an infectious person
in relation to a population in which no person is immune to the pathogen (susceptible population = total population).**

The quotient between the input and output rate of the group I, i.e., between the two terms on the right-hand side of (6.9), is

$$R(S) = R_0 \frac{S}{N} \tag{6.13}$$

and is called the **reproduction number**, sometimes also called net reproduction number. As the S value decreases, the reproduction number also decreases from the maximum value R_0.

To prevent an epidemic from spreading further, the following must apply:

$$\dot{I} = I \cdot \left(\frac{a}{N} \cdot S - b \right) \leq 0 \quad \Longleftrightarrow \quad S \leq \frac{b}{a} \cdot N = \frac{N}{R_0}.$$

So if S drops below the threshold, the virus will stop spreading; in other words, if the reproduction number $R(S) \leq 1$.

In the case $R_0 = 2$, the number of infected people decreases only if S has decreased to half the population.

At the threshold value $S = S_{max} = \frac{N}{R_0}$, the number of infected people is at its maximum. Its calculation will be discussed later.

During the initial phase of an epidemic, $S(t) \approx N$ is practically constant. Therefore, (6.9) leads to $\dot{I} = (a\frac{S}{N} - b) \cdot I \approx (a - b) \cdot I$ and hence, together with the initial condition $I(0) = I_0$, to the solution

$$I(t) \approx I_0 \cdot e^{(a-b) \cdot t}.$$

At the beginning of the epidemic I(t) grows exponentially.

---

[12] Reproduction numbers are noted in the sans serif font and are to be clearly distinguished from the function R and its initial value $R(0) = R_0$.

Example 6.45 [13]

Let $N = 8.33$ million (population of Switzerland in 2020), the two parameters $a = 0.8$, $b = 0.5$, the basic reproduction number $R_0 = 1.6$ and the initial conditions $S_0 = N$, $I_0 = 1$, $R_0 = 0$. The following diagram, where time is measured in days, displays the numerical solution of the system of differential equations:

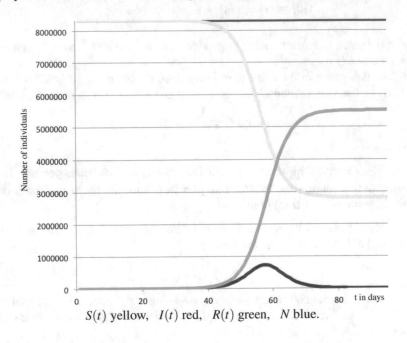

$S(t)$ yellow, $I(t)$ red, $R(t)$ green, N blue.

$\diamondsuit$

The (S, I) trajectory and the maximum number of infected people

Dividing (6.9) by (6.8) and using (6.13) implies

$$\frac{dI}{dS} = \frac{N}{R_0} \cdot \frac{1}{S} - 1. \tag{6.14}$$

In the following we consider I as a function of the variable S, knowing full well that we should actually introduce a new function to distinguish it from the earlier function $I(t)$. For the sake of simplicity, however, we refrain from it.

The maximum number of infected people occurs at the zero $S = \frac{N}{R_0} = S_{max}$ of the right-hand side of (6.14) and thus at the threshold value.

Integrating this separable differential equation with respect to the variable S leads to

$$I(S) = \frac{N}{R_0} \cdot \ln S - S + C.$$

---

[13] This and the next example are taken from [16] with the consent of the author.

Taking $I(S_0) \approx 0$ and $S_0 \approx N$, we obtain $0 \approx \frac{N}{R_0} \cdot \ln N - N + C$. Substituting C into $I(S)$ yields the trajectory

$$I(S) \approx \frac{N}{R_0} \cdot \ln S - S + N - \frac{N}{R_0} \cdot \ln N = \frac{N}{R_0} \cdot \ln\left(\frac{S}{N}\right) + N - S. \tag{6.15}$$

For $S = S_{\max} = \frac{N}{R_0}$ we obtain the maximum number $I_{\max}$ of infected people:

$$I_{\max} = \left[1 - \frac{1}{R_0} \cdot (1 + \ln R_0)\right] \cdot N. \tag{6.16}$$

Example 6.46 The parameters and initial conditions are the same as in the first example with the exception of $b = 0.4$ and

$$R_0 = \frac{a}{b} = \frac{0.8}{0.4} = 2.$$

Then $S_{\max} = N/2 = 4.16$ million and Equation (6.16) yields the numerical result $I_{\max} = 0.154 \cdot N = 1.28$ million, as evidenced by the red graph of the (S, I) trajectory (6.15), with the I-axis pointing up:

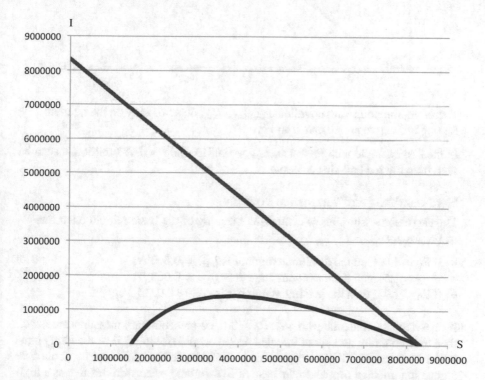

For comparison: In the first example $R_0 = 1.6$ was smaller than here, and so was $I_{\max}$, as can be seen from the earlier red curve. ◇

The (S, I) trajectory in red can be regarded as a parametrized curve that varies with time t as a parameter, starting at $t = 0$ at $S_0 \approx N$ on the right and ending at time $t = \infty$ at S_∞ on the left of the horizontal coordinate axis. As $I + S \leq N$, it is located below the straight line $I + S = N$ in blue.

For the relative maximum fraction $i_{max} = I_{max}/N$ we have

$$i_{max} = 1 - \frac{1}{R_0} \cdot (1 + \ln R_0) \qquad (6.17)$$

with the graph of i_{max} plotted upwards:

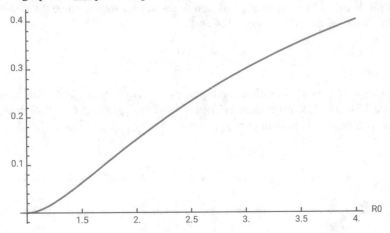

Therefore, the point with coordinates (s_{max}, i_{max}) depends only on the quotient $R_0 = \frac{a}{b}$ of the two parameters a and b.

In the first example with $R_0 = 1.6$, $I_{max} = 0.0813 \cdot 8.33 = 0.677$ million, as can be seen from the red bell-shaped curve.

Herd Immunity and Vaccination Strategy

The curve above shows how central the meaning of the basic reproduction rate R_0 is. For instance,

- if $R_0 = 3.0$, then the maximum number is $I_{max} = 0.300\ N$;
- if $R_0 = 1.5$, then the maximum number is $I_{max} = 0.063\ N$;
- if $R_0 = 1.25$, then the maximum number is $I_{max} \approx 0.0215\ N$.

If, in a rampant epidemic with, say, $R_0 = 3.0$, no precautionary measures are taken, then the maximum number of infected people would rise to 30 % of the total population. During the COVID-19 epidemic in 2020 the deaths accounted for about 5 % among the infected people. In the case of Switzerland or Austria this means a high number of around 120,000 deaths, which is ethically unacceptable. In Germany we would see around ten times as many deaths.[14]

---

[14] These values are also valid in the SEIR model, as will be shown later.

Such an approach corresponds to the strategy of achieving a so-called **herd immunity**: No precautionary measures are taken according to the motto that the population immunizes itself. Relevant political authorities of some countries used this fatal strategy even a number of days after the outbreak!

In contrast, the approach of lockdown with drastic precautionary measures, such as social distancing, thorough hand washing, closing of schools and events, quarantine measures for high risk people significantly reduces the basic reproduction rate and thus I_{max}.

If there is a **vaccine** against the disease, herd immunity can be achieved by ensuring that the infection rate decreases steadily. The right-hand side of the second differential equation (6.9) must therefore be negative. This is the case if and only if $S < \frac{N}{R_0}$. Realization: At least

$$N - \frac{N}{R_0} = \left(1 - \frac{1}{R_0}\right) \cdot N \quad \text{people must be vaccinated.} \qquad (6.18)$$

If , say $R_0 = 1.5$, then about a third of the population needs to be vaccinated.

Applications of the SIR model:

The SIR model is suitable for epidemics in which the incubation time is extremely short, such as in measles, mumps and rubella.

In the case of the Bombay epidemic (1905–1906) and a flu epidemic in 1998 at some large English boarding school, it was possible to confirm the usefulness of the SIR model. With a suitable choice of parameters a and b, the calculated graph of $I(t)$ fitted well with the data acquired from the infected individuals.

More on this can be found in [30].

6.2.2 The SEIR Model

The SEIR model is suitable for studying the behavior of the coronavirus pandemic, which is discussed in the next sections.

It is an extension of the SIR model and has an additional fourth group, named E, consisting of the *exposed individuals* who are infected, but can only infect other people after an incubation period (pre-infection period) T_p. Furthermore, let T_i be the infection time.

$$\boxed{S} \longrightarrow \boxed{E} \xrightarrow{T_p} \boxed{I} \xrightarrow{T_i} \boxed{R}$$

The SEIR model is described by the following system of nonlinear differential equations:

$$\dot{S} = -a \cdot \frac{I}{N} \cdot S \tag{6.19}$$

$$\dot{E} = a \cdot \frac{I}{N} \cdot S - c \cdot E \tag{6.20}$$

$$\dot{I} = c \cdot E - b \cdot I \tag{6.21}$$

$$\dot{R} = b \cdot I \tag{6.22}$$

Here $c = \frac{1}{T_p}$ with the pre-infectious period T_p, and $b = \frac{1}{T_i}$ with the infectious period T_i.

As (6.11) and (6.12) are still valid, the basic reproduction number R_0 and its interpretation remain the same as in the SIR model.

The sum of all right-hand sides is again zero. Therefore,

$$S(t) + E(t) + I(t) + R(t) = N.$$

The model assumes that an individual who has become immune remains immune during the epidemic. However, this assumption is not fulfilled in the case of the coronavirus epidemic. There have been known cases of immune individuals reinfecting people. Nevertheless, as long as these are isolated incidents, the model can be used.

6.2.3 Computing S_∞ and R_∞ for the Two Models.

We wish to find the number of people S_∞ for $t = \infty$ who were not infected by the epidemic. Dividing (6.19) by (6.22) or (6.8) by (6.10) and using (6.13) yields for both models

$$\frac{dS}{dR} = -\frac{R_0}{N} \cdot S. \tag{6.23}$$

In the following we consider S as a function of the variable R, knowing full well that we should actually introduce a new function to distinguish it from the earlier function $S(t)$. For the sake of simplicity, however, we refrain from it.

Integrating S with respect to the variable R and using the initial condition $S(R = 0) \approx N$ results in the (R, S) trajectory

$$S(R) = N \cdot \exp\left(-\frac{R_0}{N} \cdot R\right).$$

Since $I_\infty \approx 0$ for $t = \infty$ after the disappearance of the epidemic and $E_\infty \approx 0$ in the case of the SEIR model, the following relationship applies to both models:

$$S_\infty + R_\infty \approx N. \tag{6.24}$$

Therefore, the equation, which cannot be solved explicitly, provides

$$S_\infty = N \cdot \exp\left(-\frac{R_\infty}{N/R_0}\right) = N \cdot \exp\left(\frac{S_\infty - N}{N/R_0}\right). \tag{6.25}$$

Example 6.47 As in the first example, $N = 8.33$ million and $R_0 = 1.6$. Therefore, $N/R_0 = 5.21$ million. A Solve command (for example using Wolfram Alpha) returns for

$$S_\infty = 8.33 \cdot \exp\left(\frac{S_\infty - 8.33}{5.21}\right)$$

the solution $S_\infty = 2,988$ million, in accordance with the yellow curve.

After canceling the factor N in the exponent of Equation (6.25), we obtain an implicit equation for the relative number $s_\infty = S_\infty/N$ with parameter R_0:

$$s_\infty = \exp[R_0 \cdot (s_\infty - 1)]. \tag{6.26}$$

The following diagram shows s_∞ as a function of $R_0 \in [1,4]$:

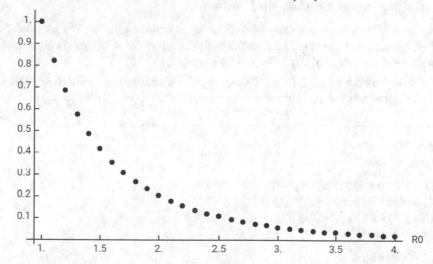

The second wave: If, after an epidemic has subsided, there is a remaining number of infected people, this can trigger another wave. Its strength depends on the number of remaining infected people S_∞.

Due to (6.24), the new initial conditions become

$$S_{new} \approx S_\infty, \quad R_{new} \approx N - S_\infty$$

where I_{new} and E_{new} (in the case of the SEIR model) are comparatively small.

Even with the assumption of 80,000 recovered people in Switzerland, the initial conditions for another wave would practically not change, since the proportion R_{new}/N of immune people would approximately be just 1 %. The same applies to other countries and regions.

The situation for a second wave would be better only if there was a high level of transmission, which unfortunately goes with a high number of deaths beforehand.

6.2.4 SEIR Model for the Coronavirus Epidemic in Germany

This section was inspired by the article [40] of the German Society of Epidemiology, March 24, 2020. The SEIR model was used with the following parameters[15]:

- Population of Germany: $N = 81.5$ million,
- $a = R_0 \cdot b = \frac{R_0}{3}$,
- $b = \frac{1}{3}$, as the infectious time span is $T_i = 3$ days,
- $c = \frac{1}{5.5}$, as the pre-infectious time span is $T_p = 5.5$ days,
- Initial conditions: $E(0) = 40{,}000$, $I(0) = 10{,}000$, $S(0) = N - E(0) \approx N$, $R(0) = 0$.

A. Computing and Graphing the Functions

Our goal is to solve the system of differential equations utilizing the parameters given above. The following Mathematica code computes the solutions via numerical integration and subsequently generates their graphs.

The four functions S, E, I, R are denoted by the corresponding lowercase letters s, e, i, r, gli stands for the i-th equation and ani for the i-th initial condition:

```
R0 = 3.0;
N = 81500000;

gl1 = s'[t] == -(R0/3N)*i[t]*s[t];
gl2 = e'[t] == (R0/3N)*i[t]*s[t]-e[t]/5.5;
gl3 = i'[t] == e[t]/5.5-i[t]/3.0;
gl4 = r'[t] == i[t]/3.0;

an1 = s[0] == N;
an2 = e[0] == 40000;
an3 = i[0] == 10000;
an4 = r[0] == 0;

sol = Flatten[NDSolve[{gl1, gl2, gl3, gl4, an1, an2, an3, an4},
 {s[t], e[t], i[t], r[t] }, {t,0,300} ]]

{ss[t_], ee[t_], ii[t_], rr[t_]}=s[t], e[t], i[t], r[t]}/.sol;

Plot[{ii[t], ee[t] }, {t,0,300}, PlotRange → All]

Plot[{ss[t], rr[t] }, {t,0,300}, PlotRange → All]
```

---

[15] I owe the precise information to Dr. Veronika Jaeger from the Westphalian Wilhelms University, Institute of Epidemiology.

For the given basic reproduction numbers $R_0 = 3.0, 1.5, 1.25$, the code delivers in each case

- a first graph with $I(t)$ as a blue and $E(t)$ as an orange curve,
- a second graph with $S(t)$ as a blue and $R(t)$ as an orange curve.

The rulings on the time axes show the number t of days after the beginning of the epidemic at $t = 0$.

Comments and comparisons of the diagrams:

(a) In the SEIR model **the function $I(t)$ also increases exponentially at the beginning** (without proof), as is visualized by the graph.

(b) Evidently, in all three cases there is a delay of $T_p = 5.5$ days of I_{max} (blue curve) against E_{max} (orange curve).

(c) Lowering R_0 has the effect of reducing I_{max}, E_{max}, and R_∞, and extending the duration of the epidemic.

(d) Solving Equation (6.26) for $R_0 = 3, 1.5, 1.25$ results in the relative s_∞-values 0.0595, 0.417, 0.629. Multiplying by $N = 81.5$ million results in the absolute S_∞-values 4.85, 34.0, 51.3 in millions. Their correctness is confirmed by the three S curves in blue.

(e) In all three cases, $S_\infty + R_\infty \approx N$ is confirmed.

(f) In all graphs, time $t = 0$ refers to March 15, 2020.

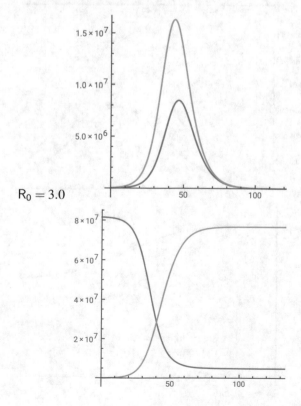

$R_0 = 3.0$

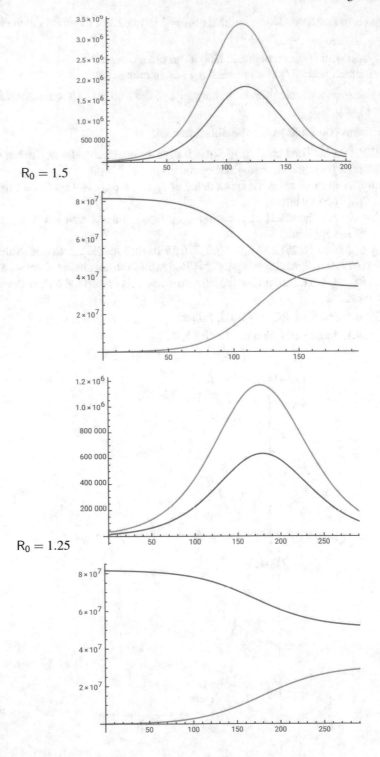

B. Total Number of Infected Individuals

For the total number of infected individuals $J(t)$ at time t we have
$$J(t) = E(t - T_p) + I(t),$$
because on average, people from group E switch to group I after a pre-infectious period T_p.

In the following, we shall work with approximations, as the time difference T_p between the maximum points of the two functions $I(t)$ and $E(t)$ is relatively small:
$$J(t) \approx E(t) + I(t) \quad \Rightarrow \quad J_{max} \approx I_{max} + E_{max}.$$

The second and third differential equations (6.20) and (6.21) imply
$$\dot{J} \approx \dot{E} + \dot{I} \approx \left(\frac{a}{N}S - b\right) \cdot I. \tag{6.27}$$

The maximum value J_{max} occurs where $\dot{J} = 0$, that is, as in the SIR model at
$$S \approx S_{max} = \frac{b}{a} \cdot N = \frac{N}{R_0}.$$

Dividing (6.27) by (6.19) yields the differential equation for the (J, S) trajectory:
$$\frac{dJ}{dS} \approx N \cdot \frac{b}{a} \cdot \frac{1}{S} - 1 = \frac{N}{R_0} \cdot \frac{1}{S} - 1.$$

$J(S)$ obeys the same differential equation (6.14) as $I(S)$ does in the SIR model. **Therefore, the relative maximum fraction $j_{max} = J_{max}/N$ still obeys Equation (6.17) and subsequently its graph is still approximately valid:**
$$j_{max} = 1 - \frac{1}{R_0} \cdot (1 + \ln R_0).$$

Let t_{max} be the approximate time of the maximum values of E, I, and J. Then
$$\dot{I}(t_{max}) \approx 0 = c \cdot E_{max} - b \cdot I_{max}.$$

Hence the two properties hold:
$$\frac{E_{max}}{I_{max}} \approx \frac{b}{c} \approx \frac{T_p}{T_i}, \qquad J_{max} \approx \left(\frac{T_p}{T_i} + 1\right) \cdot I_{max}.$$

The ratio above makes sense, because the transition from group E to group I takes T_p days, while that from group I to group R takes T_i days.

Both properties can be confirmed by measuring the graphs within measurement accuracy. Let us take a closer look at

Case $R_0 = 1.5$:

Measurements in millions: $E_{max} = 3.35$, $I_{max} = 1.82$.

So $\frac{3.35}{1.82} = 1.84$ is comparable to $\frac{5.5}{3} = 1.83$, and $(3.35 + 1.82)/81.5 = 0.0634$ is comparable to the calculated value $j_{max} = 0.0630$.

It is plausible that the maximum number of infected people is the same in both models.

A **strategy of herd immunity** in the case of $R_0 = 3.0$ with $j_{max} = 0.300$ and a death rate of an estimated 5 % would result in ethically unjustifiable 1.2 million deaths in Germany.

Numerical experiments show that I_{max} and E_{max} and thus also J_{max} with the values $R_0 = 0$, $S_0 = N$ remain practically constant with respect to changes in the initial conditions I_0 and E_0. This is true as long as the changes are relatively small compared to the total population N.

C. Possible Vaccination Strategy
Assuming that a vaccine is available, the question arises of how many people, given the R_0 value, have to be vaccinated in order for the epidemic of a population of size N to subside.

This happens if $J < 0$, i.e. if it is ensured that $S < N/R_0$ is implemented. Thus, the statement made in (6.18) applies in both the SIR and the SEIR model, which makes perfectly sense.

D. Diagrams published by the German Society of Epidemiology
They were taken from [40] and generated by using the SEIR model with the same parameters.

Diagram D1: It shows the infection curves $I(t)$. They comply with those generated earlier by Mathematica in Section A.

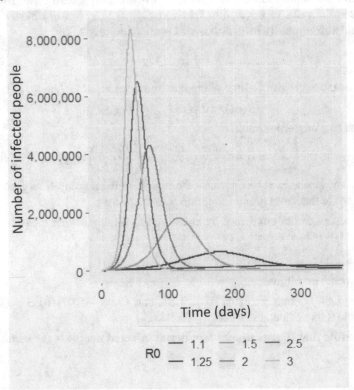

Diagram D2: It shows the evolution of the epidemic if 7, 14, 21, 28, 30 days into the epidemic, the initial reproduction number $R_0 = 2.0$ is reduced to $R_0 = 0.9$ by arranging drastic safety measures. Evidently, rapid intervention significantly reduces the maximum value:

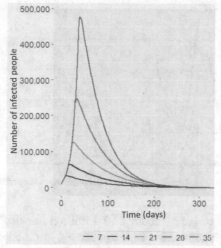

It should be noted that the scale designations must be multiplied by a factor of $T_p/T_i + 1 = 5.5/3 + 1 = 2.83$ for the curves to describe the total number $J(t)$ of infected people.

E. Diagram of the Robert Koch Institute
The following diagram, generated by Dr. an der Heiden at the Robert Koch Institute was taken from [1]. It is also based on the SEIR-model.

Since no pre-existing immunity is known, we distinguish the following two cases:

E1. Population lacking immunity: This case is represented by the three solid curves in the diagram: The red curve shows the case $R_0 = 2.0 =$ constant. The two seasonal cases were modeled by an annually periodic sinusoidal time-dependent function $R_0(t)$ with a minimum value of $\frac{2}{3} \cdot 2 = 1.33$ (slight seasonality, blue curve in the middle) or $\frac{1}{3} \cdot 2 = 0.67$ (clear seasonality, green curve on the right) in early July, and a maximum value of 2.0 at the beginning of the year. They take into account the seasonal contact behavior and the climatic conditions for the spread of the virus.

E2. Population with one third of people immunized: This case is represented by the three dashed curves in the diagram. Compared to Case E1, the R_0 values here are multiplied by a factor of 1.5: The red curve shows the case $R_0 = 3.0 =$ constant. In the case of slight seasonality (blue curve in the middle) R_0 fluctuates between 2.0 and 3.0. In the case of clear seasonality (green curve on the right) R_0 fluctuates between 1.0 and 3.0.

At the beginning there will be two secondary cases on average for $R_0 = 3.0$, as in Case E1 with $R_0 = 2.0$, since one out of three people is immune. The beginning of the epidemic is rather similar. But thereafter, the sick people will encounter more and more immune people, thus infecting fewer people.

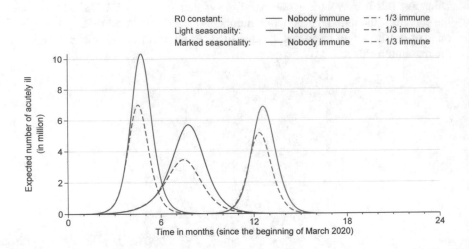

Note: The maximum of approximately 10.4 million infected people for $R_0 = 2.0$ confirms the considerations made earlier when $J_{max} \approx 2.83 \cdot I_{max}$ with $I_{max} \approx 4.4$ million was read from the blue curve in Diagram D1. The quotient $10.4/4.4 = 2.36$ deviates slightly from 2.83 because the Robert Koch Institute used slightly different parameters T_i and T_p for their calculations.

6.2.5 Coronavirus Data and SEIR Model in the First Wave

The following diagram by the *Federal Office of Public Health in Switzerland (Bundesamt für Gesundheit BAG)* with the values in thousands shows the typical evolution of the current infection numbers. The height of the orange portion of the ordinate corresponds to the graph of $I(t)$. The sum of the growing number of recovered people in purple and the number of dead people in black corresponds to the function $R(t)$.

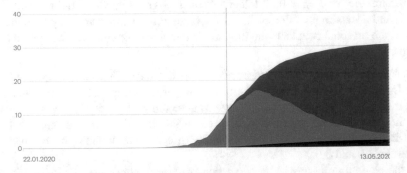

Confirmed coronavirus cases in Switzerland in thousands.
Time period: January 20 through May 13, 2020.
Vertical line: time when data source changed from Johns Hopkins University to Switzerland.

The **number of reproductions** generally depends on both time and place, be it within a country or, when comparing mean values, between different nations. Different cultural and strategic behaviors, seasonal influences and infrastructures have a crucial influence on them. The discipline in implementing precautionary measures such as hand washing, social distancing and wearing face masks plays an essential role.

In Switzerland the number of reproductions fell from the mean value of 1.8 to 0.7 in the period from March 7, 2020 to April 27, 2020, partly due to the introduction of lockdown measures in the period from March 13 to 16, 2020.

The Huazhong University of Science and Technology in Wuhan, China found in a study that in the 11-million city of Hunan R_0 was 3.18 in the first week of January 2020, and after drastic security measures dropped to just 0.32 in February.

The following diagram (published in *Neue Zürcher Zeitung NZZ* on April 9, 2020) using a logarithmic scale **compares Cumulative Infection Curves of severely affected regions.** They correspond to the graphs of the functions $R(t)$, with their limit values R_∞ equaling the number of people who have become immune or who have died.

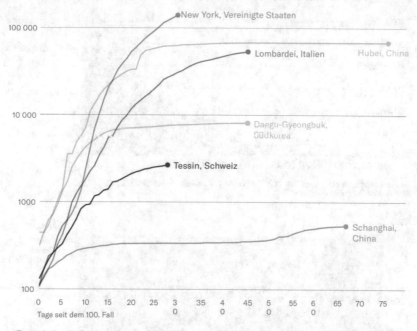

Cumulative Infection Curves from February 25 (mark 0) to April 13, 2020 (mark 45)

Sources: Canton of Zurich, Johns Hopkins University, Italian Ministry of Health, Korea Centers for Disease Control and Prevention, New York Times

The Canton Tessin (Switzerland) appears in the above diagram separately, even though it has just 354,000 inhabitants. The reason is that the epidemic in Switzerland began there as it borders on the Bergamo region in northern Italy which was severely hit by the pandemic. Populations in millions: New York 20, Hubei 58.5, Lombardy (Italy) 10, Daegu (South Korea) 2.5, Shanghai 26.

According to (6.22) and (6.10) and using $b = 1/T_i$ for the two models,

$$R(t) = \frac{1}{T_i} \cdot \int_0^t I(\tau) \cdot d\tau.$$

Dividing the integral by T_i days accounts for the fact that every infected person is counted for an average of T_i consecutive days, i.e. T_i times.

$I(t)$ increases exponentially at the beginning and so does the integral, i.e., the function $R(t)$.

At the beginning, the curves are approximately linear, thus confirming the exponential growth of the functions $I(t)$ mentioned previously. The reason is that on a logarithmic scale exponential functions appear as straight lines.

6.2.6 World Data in the Second or Third Wave

Absolute:

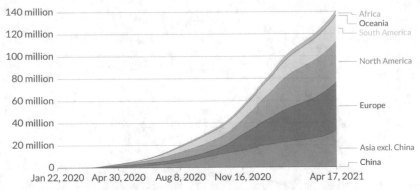

Source: Johns Hopkins University CSSE COVID-19 Data – Last updated 18 April, 10:02 (London time)
OurWorldInData.org/coronavirus • CC BY

Relative:

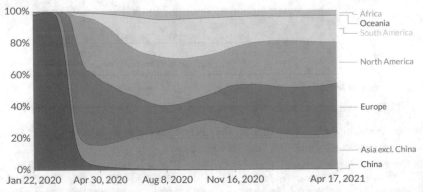

Source: Johns Hopkins University CSSE COVID-19 Data – Last updated 18 April, 10:02 (London time)
OurWorldInData.org/coronavirus • CC BY

6.3 Exercises Chapter 6

97. Climate Data.

(a) Using the appropriate graph, verify that the present CO_2 concentration is about 30 % higher than all the values in the past 800,000 years.

(b) The calculated c_2 value for the 2.0 degree target is $c_2 = 491$ ppm. What is its relation with the graph that shows data of the paleoclimate for 100 million years and the increased mean earth's surface temperature of $14.7°C + 2.0°C$?

(c) For the two given global data $\Delta^{14}C \approx 19\%$ (1987) and $\Delta^{14}C \approx 2\%$ (2017), use Equation (6.7) to compute the ratios $\dfrac{n^{14}}{n^C}$ and compare the two.

98. Carbon Budget for the 1.5 Degree Target.
Compute the carbon budget for the target temperature of $1.5°$ C in a manner analogous to that in the last section. It should be taken into account that the non-CO_2 forcings are responsible for 10 % of the temperature increase of $1.5°$ C.

How likely is it that the effective temperature increase is less than $1.5°$ C?

99. Time Dependent Radiation Function g(t).
For differing target temperatures Δ_∞, compute the temperature curve of $\Delta(t)$ for various scenarios by using time-dependent radiation functions $g(t)$ that are chosen realistically.

100. SIR Model.

(a) Verify the following:

  (i) With increasing infection parameter a and fixed immunity parameter b, the maximum number of infected people I_{max} increases and S_∞ decreases.

  (ii) With increasing immunity parameter b and fixed infection parameter a, the maximum number of infected people I_{max} decreases and S_∞ increases.

  (iii) If both parameters a and b are multiplied by the same factor, the quantities I_{max} and S_∞ remain unchanged. However, the epidemic spreads faster and also subsides faster. The bell-shaped graph of $I(t)$ becomes narrower and occurs earlier in time. The vertex of the parametrized curve $I(S)$, however, remains unchanged.

(b) It has been shown that the implicit formula for s_∞ applies to both models. Derive the formula again using the (S, I) trajectory. Note: $I_\infty = 0$.

101. Vaccination Campaign.
In a population of N people, an epidemic breaks out with the basic reproduction number $R_0 = 3$. This means that each infected person infects on average three other people.

As a vaccine is available, vaccination will be administered after the outbreak of the disease. Show that vaccinating about $2/3$ of the population is enough to reduce the number of infected people and therefore prevent the disease from spreading.

102. SEIR Model.

(a) The total population N and S_∞ are known. Find $R_\infty, I_\infty, E_\infty$.

(b) In the case of basic reproduction number $R_0 = 2.0$, determine the quantities S_∞, R_∞ and the maximum total number of infected people J_{max} for a population of size N.

(c) It is known that a reproduction number of $R = 0.91$ results in halving the number of infected people in 5 months. For which R-value is the halving time 8 weeks? (By U. Burri.)

(d) Implement the SEIR model for the coronavirus epidemic. Experiment with parameters and initial conditions and generate the diagrams of the four functions in the same way as was done in the case of Germany.

103. Second Wave. In a population of size N the number of people remaining susceptible $S_\infty = \lambda \cdot N$ was determined from the existing data. Let the value of R_0 be known.
As was shown, a second wave occurs if $\dot{J}(0) > 0$, i.e. if the inequality $\lambda N > N/R_0$ and thus $\lambda \cdot R_0 > 1$ is satisfied.

(a) Using the function $J(S)$ with the corresponding initial condition, show that after the second wave the relative maximum fraction of infected people is

$$j_{max} = \lambda - \frac{1}{R_0} \cdot (1 + \ln(\lambda R_0)).$$

(b) Using the (R, S) trajectory, show that after the second wave the following implicit equation for s_∞ holds:

$$s_\infty = \lambda \cdot \exp[-R_0(\lambda - s_\infty)].$$

(c) Consider a certain population with R_0 in the first wave and R_0^* in the second wave.

   (i) For a **constant risk of infection** in the second wave, the value is $R_0^* = R_0/\lambda$, because only λN people can be infected.
   Compute for both waves j_{max} and s_∞, if $R_0 = 1.25$.

   (ii) For a **higher risk of infection** in the second wave, $R_0^* > R_0/\lambda$.
   Use $R_0 = 1.25$ and $R_0^* = 2.50$ to compute the values j_{max} and s_∞ for the second wave. Verify that the second wave is much stronger than the first wave.

(d) Prove the following: If R_0 remains the same or is smaller after the epidemic, there is no second wave, no matter how large R_0 is.

Hint: For a fictitious second wave, consider the solution s_∞ of the implicit equation as the intersection of the graphs of x and $f(x) = \exp[-R_0(1 - x)]$ at $x = \lambda$. Show that the slope at this point is $f'(\lambda) < 1$ and compute the slope.

Chapter 7
Solutions

7.1 Solutions Chapter 1

1 Exponential Functions.
$b^x = e^{(\ln b)\cdot x}$, etc.

2 Exponential Function Through a Given Point.
Plug in the coordinates of the given point and determine C.
  Check your answer by plotting.

3 Uranium.
  (a) 1602 years, 3.824 days.
  (b) both are 0.
  (c) $\dot{m} = 0 \Rightarrow \dfrac{\lambda_2}{\lambda_1} = \dfrac{\exp(-\lambda_1 t)}{\exp(-\lambda_2 t)} \rightarrow$ in $t = 0.181$ years, i.c. 66 days.

4 Brainteaser: Sprinter and Tortoise.
The time T until the sprinter passes the tortoise is a geometric series with $q = 0.1$,
  which converges toward
$$T = 1 + q + q^2 + q^3 + \ldots = \frac{1}{1-0.1} = \frac{10}{9}.$$

In $1\frac{1}{9}$ s and at distance $11\frac{1}{9}$ m the sprinter overtakes the tortoise.

5 Gaußian Bell Curve.
The graph has a line symmetry in the y-axis and passes through point $(0, 1)$.
  The inflection points are $(x_W, y_W) = \left(\pm\dfrac{1}{\sqrt{2k}}, \dfrac{1}{\sqrt{e}} \right)$. The x-axis is an asymptote.

7 Logarithms.
(a) $a \approx 0.4343$,   (b) Substitute $u = b^x, v = b^y$.

8 Growth Model for Young Children.
(a) 86.8 cm, 9.7 cm/year,   (b) at $\frac{1}{4}$ year = three months.

© The Author(s), under exclusive license to Springer Nature Switzerland AG 2021
A. Fässler, *Fast Track to Differential Equations*,
https://doi.org/10.1007/978-3-030-83450-0_7

10 Least Square Method.

$x = \frac{1}{n} \cdot (x_1 + x_2 + \ldots + x_n)$ is the arithmetic mean of the measured data.

11 Puzzling Worm's Eye View.

$$D(h) = \sqrt{h(2R+h)} \approx \sqrt{2R} \cdot \sqrt{h}.$$

The graph has a vertical tangent at $h = 0$.

If $h = 1$ cm, then $D = 357$ m.

For $h = 100$ m, we have 35.7 km.

And $h = 1$ mm gives a surprising 113 m!

12 Maximum Consumer Power.

$R = R_i$.

13 Optimal Packaging.

$1 = \pi r^2 h$ surface area $O(r) = 2\pi r^2 + 2\pi r \cdot \dfrac{1}{\pi r^2}$, $r_{opt} = \sqrt[3]{\dfrac{1}{2\pi}}$.

Diameter : height = 1 : 1. In real-life cans, base and lid are more expensive, and waste from punching circular disks also plays a role.

14 Salmon Swimming with Minimum Energy.

Energy = time × power: $W(r)$ is proportional to $\dfrac{s}{r-v} \cdot r^\lambda$.

Differentiating and equating to zero yields $r_{opt} = \dfrac{\lambda}{\lambda - 1} \cdot v$.

15 Elasticity of a Function.

(a) Parabolic segment from point $(0, 400)$ to vertex $(20, 0)$.

(b) $\varepsilon_{N,p} = \dfrac{2p}{p - 20}$.

(c) $\varepsilon_{N,5} = -\dfrac{2}{3}$, $\varepsilon_{N,19} = -38$.

If the price rises 1%, then the demand drops -0.67% or -38%, resp. The closer the price approaches 20, the faster N decreases in percentage as compared to p.

17 Brainteaser: Snowplow.

Assume that snowing starts at time $t = 0$. Snow height $h = at$. At some yet unknown time $t = x$ the snowplow starts working. As the power is constant, $v = b/t$ must be inversely proportional to h. Distance covered = integral of v:

$$\int_{x}^{x+1} v(t) \cdot dt = b \cdot \ln\frac{x+1}{x} = 2 \qquad \int_{x+1}^{x+2} v(t) \cdot dt = b \cdot \ln\frac{x+2}{x+1} = 1$$

Dividing the equations

$$\ln\frac{x+1}{x} / \ln\frac{x+2}{x+1} = 2 \implies \ln\frac{x+1}{x} = 2\ln\frac{x+2}{x+1} = \ln\left(\frac{x+2}{x+1}\right)^2 \implies \frac{x+1}{x} = \left(\frac{x+2}{x+1}\right)^2$$

yields the solution $x = (\sqrt{5} - 1)/2$, which equals the golden ratio g. At time $(8 - g)$ it started snowing. Note that the answer does not depend on the parameters a and b.

18 Brainteaser: Spider Kunigunde.

(a) Use the relative distance $D(t)$, i.e. work with percentages:

Let $\varepsilon = \dfrac{1 \text{ mm}}{1000 \text{ km}} = 10^{-9}$. Then $D(1) = \varepsilon$. The second second contributes only half of it, hence $D(2) = \varepsilon + \frac{\varepsilon}{2}$. After n seconds

$$D(n) = \varepsilon \cdot \left(1 + \frac{1}{2} + \frac{1}{3} + \ldots + \frac{1}{n}\right).$$

Expression between parentheses = harmonic sum = H_n.

As the harmonic series diverges, Kunigunde does reach the finish!

(b) For the arrival time T we have $100\% = 1 \approx \varepsilon \cdot H_T$. Due to (1.10) we have $H_T \approx \ln T$ for huge T. Hence the equation is $\ln T \approx 10^9 \Rightarrow T \approx \exp(10^9)$ s. That's a number beyond imagination having more than 10^8 digits! Astronomers estimate the number of atoms in the whole universe to be roughly 10^{85}, i.e. a number with "merely" 86 digits. This is a ridiculously small number compared with T. Nevertheless, Kunigunde wouldn't even reach this age. Remark: The harmonic series diverges extremely slowly.

20 Kepler's Barrel Rule (Newton-Cotes formula).

The integral on the left is $\quad \dfrac{ah^4}{4} + \dfrac{bh^3}{3} + \dfrac{ch^2}{2} + dh.$

Show that the right-hand side provides the same answer.

21 Checking Integral.

The answer describes exponentially damped oscillations.

22 Geometry of Railway Tracks, Roads and Ski Jumps.

(a) Segment $y = 0$ joined at $x = 0$ to $y = ax^3$ has the second derivatives $y'' = 0$ joined to $y'' = 6ax$. The overall function y'' is piecewise linear with a cusp at $x = 0$, hence continuous.

(b) (i) $K(\ell) = \dfrac{\sqrt{\pi}}{A}\ell.$

(iii) $v = A\sqrt{\pi} = \text{const}, L(\ell) \cdot R(\ell) = A^2 = \text{const}.$

(iv) $\frac{v^2}{R} = 0.7g, R = 112.4m.$

23 Piecewise Linear Functions.

(a) $G(x) = \begin{cases} \frac{1}{2}x^2 - \frac{1}{2} & \text{if } x \leq 1 \\ x - 1 & \text{if } 1 < x < 2 \\ -(x-3)^2 + 2 & \text{if } x \geq 2 \end{cases}.$

Parabola with vertex $(0, -1/2)$ followed by a straight line segment with slope 1 and by another parabola with vertex $(3, 2)$. Smooth at $x = 1$, cusp at $x = 2$ with slopes 1 and 2.

(b) $f'(x) = \begin{cases} 1 & \text{if } x < 1 \\ -1 & \text{if } x > 1 \end{cases} \quad$ discontinuous at $x = 1$.

24 Brainteaser: Express Train and Continuity.
Let $s(t)$ be the distance traveled by the train during the time interval $[t,\ t+1]$. The function is continuous and defined for $0 \le t \le 1$. Moreover, $s(0) + s(1) = 200$.
Due to the Intermediate Value Theorem for functions (a continuous function $f(x)$ on a closed interval $a \le x \le b$ takes on any value between $f(a)$ and $f(b)$ at least once), there exists a t_0 such that $s(t_0) = 100$.
During the time period $[t_0,\ t_0+1]$ the train travels exactly 100 km.

25 Concentric Circle Method for an Ellipse and Paramerizing a Hyperbola.

(a) Expressing the coordinates of P in terms of a, b, t results in the parametric equation of the ellipse.
(b) Use $\cosh^2 t - \sinh^2 t = 1$.

26 Lissajous Figure.
(a) Figure-eight curve (8).
$$t = k \cdot \tfrac{\pi}{4},\ k = 0,1,\dots,8 \text{ and } (x,y) = (\pm 1, \pm\tfrac{1}{\sqrt{2}}),\ (0,\pm 1)$$
(b) Intersection at $(0,0)$ with angle $= 2\arctan\tfrac{1}{2}$.

(c) $\dot{\vec{r}}(\tfrac{\pi}{2}) = \begin{pmatrix} -2 \\ 0 \end{pmatrix}$ is horizontal.

27 Controlling a Milling Cutter.
$$\vec{r}_P = \begin{pmatrix} t \\ at^2 \end{pmatrix},\quad \vec{f} \perp \dot{\vec{r}}_P = \begin{pmatrix} 1 \\ 2at \end{pmatrix},\quad \vec{f} = \frac{R}{\sqrt{1+4a^2t^2}} \begin{pmatrix} -2at \\ 1 \end{pmatrix},$$

$$\vec{r}_F(t) = \vec{r}_P(t) + \vec{f}(t) = \begin{pmatrix} t - \dfrac{2aRt}{\sqrt{1+4a^2t^2}} \\ at^2 + \dfrac{R}{\sqrt{1+4a^2t^2}} \end{pmatrix}.$$

The parallel curve is not a parabola.

28 Self-similarity of the Logarithmic Sprial Revisited.
To simplify computations, note that for computing an angle the lengths of the vectors in the dot product are inessential. The angle φ results from $\cos\varphi = \dfrac{a}{\sqrt{1+a^2}}$.

29 Pedal Curve.

(a) (i)
$$\begin{pmatrix} x(t) \\ y(t) \end{pmatrix} = \begin{pmatrix} t - r\sin t \\ 1 - r\cos t \end{pmatrix}$$

(ii)
$$\vec{v} = \begin{pmatrix} 1 \pm r \\ 0 \end{pmatrix}$$

(b) (i)

$$\vec{r}_p(t) = \begin{pmatrix} t - \frac{1}{2}\sin(\omega t) \\ \frac{1}{2} - \frac{1}{2}\cos(\omega t) \end{pmatrix} \quad \text{with period } \frac{2\pi}{\omega}$$

(ii)

$$\vec{v}_p = \begin{pmatrix} 1 \pm \frac{\omega}{2} \\ 0 \end{pmatrix}, \quad \begin{cases} \omega > 2 \text{ if loop} \\ \omega < 2 \text{ if serpentine} \end{cases}$$

(iii)

$$\frac{1}{2}\begin{pmatrix} t - \sin t \\ 1 - \cos t \end{pmatrix} \text{ and } \frac{1}{2}\begin{pmatrix} 2t - \sin(2t) \\ 1 - \cos(2t) \end{pmatrix} \text{ describe the same cycloid.}$$

The affinity $x \mapsto \frac{1}{2}x$, $y \mapsto y$ leads to $\frac{1}{2}\begin{pmatrix} t - \frac{1}{2}\sin(2t) \\ 1 - \cos(2t) \end{pmatrix} = \vec{r}_p(t)$,

where $\omega = 2$.

30 Wheel on a Circle.

The answer may be surprising: $\vec{r}(t) = \begin{pmatrix} 4\cos t + \cos(4t) \\ 4\sin t + \sin(4t) \end{pmatrix}$

31 Brainteaser: The Rain Problem.

(a) • From the front the rain pelts with relative speed $v + r$ onto the surface. The amount of rain per second onto the surface is proportional to $v + r$. Running time is $t = s/v$. Thus the total amount of rain is proportional to

$$\frac{s}{v} \cdot (v + r) = s \cdot \left(1 + \frac{r}{v}\right).$$

Solution: Run against the rain as fast as you can!
• From the back the rain pelts with relative speed $r - v$ onto the surface.

Analogously to the previous, the total amount of rain is proportional to

$$\frac{s}{v} \cdot (r - v) = s \cdot \left(\frac{r}{v} - 1\right).$$

Solution: Run with speed $v = r$, with the amount of rain being ≈ 0!

(b) Only the component of the rain speed perpendicular to the surface contributes to the solution. Hence, in the previous solution, we just replace r by the component perpendicular to the surface.

(c) If the velocity $\vec{r}$ hits the surface element obliquely, the amount of rain per second is the same as that onto the surface projected perpendicularly to the direction of motion $\vec{v}$ with the speed component of $\vec{r}$ in the direction of motion. The latter is the same for all surface elements. The sum of all projected surface elements equals the area of the projected surface.

However, if the rain comes mainly from above, then the projection of the body from above is essential, as is the fact that the legs are moving while running.

32 Fountain.

(a) See projectile motion: $\begin{pmatrix} x(t) \\ y(t) \end{pmatrix} = \begin{pmatrix} \cos\alpha \cdot v_0 \cdot t \\ \sin\alpha \cdot v_0 \cdot t - \frac{g}{2} \cdot t^2 \end{pmatrix}$

(b) $y = -\dfrac{g}{2v_0^2 \cos^2\alpha} \cdot x^2 + \tan\alpha \cdot x$

(c) $\dot{y} = 0 \Rightarrow t_m = \dfrac{v_0 \sin\alpha}{g} \implies \begin{pmatrix} x_m \\ y_m \end{pmatrix} = \dfrac{v_0^2}{2g} \begin{pmatrix} \sin(2\alpha) \\ \sin^2\alpha \end{pmatrix}$

(d) If it is an ellipse, then with semiaxes

$$a = x_m(45°) = \frac{v_0^2}{2g}, \quad b = y_m(45°) = \frac{a}{2}, \quad \text{and center } (0,b).$$

Hence show for all α that $\dfrac{x_m(\alpha)^2}{a^2} + \dfrac{[y_m(\alpha) - b]^2}{b^2} = 1$.

With $s = \sin\alpha$ and $c = \cos\alpha$ we have

$$\frac{[a\sin(2\alpha)]^2}{a^2} + \frac{[a\sin^2\alpha - \frac{a}{2}]^2}{a^2/4} 4s^2c^2 + 4(s^4 - s^2 + \frac{1}{4}) = 4s^2(c^2 - 1) + 4s^4 + 1 = 1.$$

(e) Quadratic growth, i.e. doubling v_0 results in quadrupling the lengths.

(f) $v_0 = 4.43 \text{ m/s} = 16 \text{ km/h}$.

7.2 Solutions Chapter 2

34 Graphing Solutions with Isoclines.

(a) The solutions are equilateral hyperbolas opening left and right, or up and down, centered at the origin, with asymptotes $y = \pm x$, which are also solutions.

(b) The isoclines are concentric circles with radius $\sqrt{m}$, where m is the slope of the solution at points on the circle.

(c) The isoclines are straight lines through the origin and the solutions are spiral-shaped.

35 Nonhomogeneous Differential Equations.

(a) $y(x) = Ce^{x^2} - \dfrac{1}{2}.$ (b) For $x > 0$: $y(x) = \dfrac{C}{x} - 4$, singularity at $x = 0$.

(c) For $x > 0$: $y(x) = \dfrac{1}{x}(k - \cos x)$, singularity at $x = 0$.

36 Discontinuous Initial Value Problem..

$$y(t) = \begin{cases} e^t & \text{if } 0 \le t \le 1 \\ e & \text{if } t > 1 \end{cases}$$

37 Uniqueness.
There exists a unique solution everywhere, because all hypotheses of the Existence and Uniqueness Theorem are satisfied.

38 Singularities.
$C = ab - a^3/3$. The solution $y = x^2/3$ passing through the origin is
also unique. This is consistent with the Existence and Uniqueness Theorem, which only gives sufficient criteria.
The solution through the origin illustrates that the hypotheses of the theorem are not necessary.

39 Branching of Solutions.

(b) $B(2, -1)$

(c) Partial derivative with respect to y yields a radical expression in the denominator, which at point B has the value 0: singularity at B.

40 Infinitely Many Solutions.
$y(x) = m \cdot x$.
The origin is a singularity of the differential equation.

42 Nonlinear Initial Value Problem.

$$y(x) = \frac{2}{2 - e^{-x^2/2}}.$$

43 Stability.

(a) S-shape for $0 < y < 1$, strictly decreasing for $y < 0$ and $y > 1$.

(b) $y = 0$: unstable, $y = 1$: stable.

(c) $y(x) = \dfrac{2}{2 - e^{-x}}.$

44 Initial Value Problem.
Check your answer by substituting into the differential equation.

46 No Closed Elementary Form.
$\frac{y^2}{2} - \cos y = x + C$. Solution y cannot be written in closed elementary form.

47 Geometrical Problem.
$y(x) = C \cdot x^2$.

48 Equilibrium Points.
(a) $y_p = -b/m$ is stable if $m < 0$, it is unstable if $m > 0$.
  (b) One of the two equilibrium points is semi-stable.

7.3 Solutions Chapter 3

49 Coffee.

$$t = \frac{\ln(75/17)}{\ln(75/66)} \approx 11.6 \text{ min.}$$

50 Radium.

In general, $e^{-kT_{1|2}} = 1/2 \Longrightarrow k \cdot T_{1/2} = \ln 2$. It takes $t = 247$ years.

51 Compound Interest.

(a) $\exp\left(n \cdot \frac{p}{100}\right)$ million euros.

(b) $\left(1 + \frac{p}{100}\right)^n$ million euros.

(c) Difference is $(e^{0.3} - 1.03^{10}) \cdot 10^6 = 5,942$ euros.

(d) Continuous compounding of interest reduces the principal by $(e^{-0.3} - 0.97^{10}) \cdot 10^6 = 3,394$ euros more than annual compounding. Even in this case, continuous compounding is preferable.

52 Barometric Formula for the Density.

(a) Pressure and density are proportional to each other.

(b) They each are $\frac{1}{13,211}$ of the values at sea level.

53 Chernobyl and Vegetables.

0.55%. After two months, consumption of vegetables was rightly deemed unharmful.

54 Growth of a Cell.

$O \sim V^{2/3}, \quad r \sim V^{1/3}, \quad r(t) = kt + r_0$ is linear.

Problem similar to that of the rain drop.

55 Mixing Problem.

(a) 6 kg,

(b) 384 min and 522 min (more than two hours longer!)

56 Conical Tank.

(a) $t_E = \dfrac{\pi R^2 H}{3L}$.

(b) $h(t) = \sqrt[3]{\dfrac{3LH^2}{\pi R^2}} \cdot \sqrt[3]{t}, \quad v(t) = \dfrac{1}{3}\sqrt[3]{\dfrac{3LH^2}{\pi R^2}} \cdot t^{-2/3}$.

(c) $h(t_E) = H$, $v(0) = \infty$.   Recall Einstein's Relativity Theory!

(d) $V(h) = L \cdot t = \dfrac{\pi}{3}\left(\dfrac{R}{H}h\right)^2 h \implies t(h) = \dfrac{\pi}{3L}\dfrac{R^2}{H^2}h^3$

$w(h) = v(t(h)) = \dfrac{LH^2}{\pi R^2} \cdot \dfrac{1}{h^2} \implies \dfrac{w(H)}{w(H/2)} = \dfrac{1}{4}$.

Dimensions are consistent: $[w(H)] = \dfrac{m^5/s}{m^4} = \dfrac{m}{s}$.

Indeed, applying the laws of exponents yields $v(t_E) = \dfrac{L}{\pi R^2} = w(H)$.

(e) $t_E = 523.6$ s, $\quad v(t_E) = w(H) = 0.637$ mm/s, $\quad v(t_E/2) = 1.01$ mm,
$h(t_E/2) = 0.794$ m.

57 Smoking Problem.

(a) $\dot{V} = 0.12 - \dfrac{3}{50,000}V$ (b) After 5 min.

58 Motor Boat.

(b) 20 m/s,   (c) T = 115 s, distance = 1403 m.

59 Ball in a Gravitational Funnel.

$r(h) = R \cdot \left(\dfrac{v_0}{v(h)}\right)^{7/5}$ where $v(h) = \sqrt{v_0^2 + \dfrac{10}{7}gh}$.

The lower opening has radius $r(0.4 \text{ m}) \approx 5.33$ cm and is slightly larger than in the case of the rotating coin where radius $= 5.07$ cm.

60 Leaking Container.

$T = 500^2\sqrt{\dfrac{2}{g}}$ seconds.

Compute the numerical value yourself. A long period of time!

61 Free Fall inside a Liquid.

(b) $v(t) = \dfrac{g}{k}(1 - e^{-kt}), \qquad y(t) = \dfrac{g}{k} \cdot t - \dfrac{g}{k^2}(1 - e^{-kt})$.

(c) $m = 2.18$ g.
In glycerine $k = 389$ s$^{-1} \Rightarrow v_\infty = 2.52$ cm/s.
In olive oil it is $(15.0/1.07)$ times that of glycerine, i.e. 35.3 cm/s.

(d) It is $(7.8/19.3)$ times that of the gold ball.

62 Suspension Bridge.

$V(x + \Delta x) - V(x) = \rho g \cdot \Delta x, \qquad Hf'' = \rho g$.
Integrating twice yields a quadratic function, hence a parabola.

63 Specified Elasticity Function.

$f(x) = k \cdot x^a$.

64 Function = Elasticity Function.

Separable differential equation for f with solutions $f(x) = \dfrac{1}{k - \ln x}$.

Graph for $k = 1$:

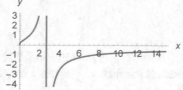

65 Tree Growth According to Chapman-Richards.

(b)

$$\dot{u} = \frac{b}{6a} - \frac{c}{6a} \cdot u \quad \Rightarrow \quad h(t) = \left[\frac{b}{c} - \left(\frac{b}{c} - \sqrt{h_0}\right) \exp\left(-\frac{c}{6a} \cdot t\right)\right]^2$$

Remark: $\frac{b}{c} > \sqrt{h_0}$

(c) $(b/c)^2$.

(d) Daily periods and in particular saisonal periods are the reasons for taking mean values for the constant parameters. Otherwise, time-dependent parameters are to be introduced with corresponding numerical solutions.

66 Logistic Growth Model.

(a) For all solutions with $p_0 > 0$, we have $\lim\limits_{t \to \infty} = \dfrac{a}{b}$.

    (i) If $0 < p_0 < \frac{a}{b}$, then the solution is an S-shaped curve, with symmetry in the point with function value $p - \frac{a}{2b}$.

    (ii) If $p_0 > \frac{a}{b}$, then the curve decreases monotonically and asymptotically towards $\frac{a}{b}$.

    (iii) If $p_0 = \frac{a}{b}$, then the solution is constant.

(b) $D > 0$, $\quad -\frac{b}{a} < D < 0$, $\quad D = 0$.

67 Chemical Reaction.

$y(t) = pq \dfrac{1 - e^{k(p-q)t}}{q - pe^{k(p-q)t}}$.

68 Models for Tumor Growth.

(a) $T = \frac{\ln 2}{\alpha}$.

(b) $V_\infty = e^{r/s} \cdot V_0$ with differential equation $V'(t) = re^{-st} \cdot V(t)$.

69 Friction of Air.

(a) $v(t) = \left(\dfrac{k}{m}t - \dfrac{1}{v_0}\right)^{-1}$.   (b) $x(t) = \dfrac{m}{k} \ln \dfrac{m + kv_0 t}{m}$.   (c) $v(\infty) = 0$, $\quad x(\infty) = \infty(!)$

70 Orthogonal Trajectories.

$y' = -\dfrac{1}{3}\dfrac{x}{y}$ describes a family of ellipses with ratio between semiaxes $\sqrt{3} : 1$.

7.4 Solutions Chapter 4

72 Mechanical Oscillator.

$$A = \sqrt{y_0^2 + \left(\frac{v_0}{\omega}\right)^2}$$

73 Toy Oscillator.

$m\ddot{y} = -Dy$, $\omega = \frac{2\pi}{T} = 6.98 \text{ s}^{-1}$, $D = m\omega^2 = 3.90 \text{ N/m} = 0.0390 \text{ N/cm}$.

A mass of about 4 g extends the spring by about 1 cm, hence a weak spring.

74 A Child's Dream.

Body weight $= G = mg$, where $g \approx 9.81 \text{ m/s}^2$, $R = $ earth's radius.

Magnitude of the gravitational force is linear in r: $|K(r)| = G \cdot \frac{r}{R}$.

Equation of motion: $\ddot{r} = -\frac{g}{R} \cdot r$. Here $-R \le r \le R$. Hence

$$\omega = \sqrt{\frac{g}{R}} \approx 0.00124 \text{ s}^{-1}, \quad T = \frac{2\pi}{\omega} \approx 1.4 \text{ h}, \quad v_{max} = R \cdot \omega \approx 7.91 \text{ km/s}.$$

76 Breakage of a Spring.

Oscillator with weak damping:

$D = 981 \text{ N/m}$, $\omega_0 \approx 9.90 \text{ s}^{-1}$, $T \approx 0.634 \text{ s}$ (approximately, since $\omega \approx \omega_0$).

$e^{-\rho T} \approx 0.97$ implies $\rho \approx 0.0480$. $A_{res} \approx \dfrac{1/10}{2 \cdot 0.0480 \cdot 9.90} = 0.105 \text{ m}$.

The spring breaks when the resonance frequency is applied to it.

The devastating effect is observed already at a modest force amplitude of 1 N.

77 Festival of Hyperbolic Functions.

(a) $\ddot{h} = \frac{g}{\ell} \cdot h$, $h(t) = h_0 \cosh\left(\sqrt{\frac{g}{\ell}}t\right)$, $v(t) = \sqrt{\frac{g}{\ell}} h_0 \sinh\left(\sqrt{\frac{g}{\ell}}t\right)$

(b) $t_{end} = \sqrt{\frac{\ell}{g} \text{arcosh}\left(\frac{\ell}{h_0}\right)}$

(c) For $h_0 = 0.5 \text{ m}$: $t_{end} = 0.690 \text{ s}$,
$$v(t_{end}) = 0.5 \cdot 2.56 \cdot \sinh(2.56 \cdot 0.69) = 3.62 \text{ m/s}.$$
For $h_0 = 0.05 \text{ m}$: $t_{end} = 1.60 \text{ s}$,
$$v(t_{end}) = 3.82 \text{ m/s}.$$

78 Magnetic Field of a Straight Conductor, Biot-Savart's Law.

$$d\ell = \frac{r}{\sin\varphi} \cdot d\varphi, \quad r = \frac{a}{\sin\varphi}, \quad 0 \le \varphi \le \pi, \quad H(a) = \frac{i}{2\pi} \cdot \frac{1}{a}.$$

79 RC Circuit.

$$R\dot{q} + \frac{q}{C} = U, \quad q(t) = CU(1 - e^{-t/RC}), \quad i(t) = \frac{U}{R} e^{-t/RC}.$$

80 Predator-Prey Model.

(a) $P(\frac{c}{d}, \frac{a}{b})$.

(b) $\left. \begin{array}{l} \dot{x} \approx -\frac{bc}{d} \cdot y \\ \dot{y} \approx \frac{ad}{b} \cdot x \end{array} \right\}$

(c) $y' = -\lambda \frac{x}{y}$ where $\lambda = \frac{a \, d^2}{c \, b^2} \implies \lambda x^2 + y^2 = C > 0$

Ellipses with semiaxes $\sqrt{C/\lambda}$ and $\sqrt{C}$. Thus ratio $= \frac{1}{\sqrt{\lambda}} = \frac{b}{d} \cdot \sqrt{\frac{c}{a}}$.

81 Coriolis Force in the Southern Hemisphere.

The analytic expressions for $\vec{w}(t)$ and $\vec{r}(t)$ given in Section 4.5 remain unchanged. But since in the southern hemisphere $f < 0$, the circular clockwise rotation must be reflected in the y-axis, resulting in a counterclockwise rotation.

82 Initial Value Problem.

$$\ddot{y} + 245y = 5\cos(3t), \quad y(0) = 0.05, \quad \dot{y}(0) = 0. \quad \text{Dimensions: m, kg, s.}$$

83 Weak Damping.

(a) $\omega_0^2 > \rho^2$: $\quad 1 > 0.0625^2$, $\quad$ (b) $\Omega_{res} = 0.996/s, \quad A_{res} = 24.0$.

85 Perfect Resonance.

(b) Natural frequency of the system $\omega_0 = \omega$, since no damping.

(c) The numerator describes a difference of two processes with almost the same frequencies; phenomenon of beats.

y varies between $\pm \dfrac{A}{\omega^2 - \Omega^2}$; resonance.

86 Beats.

$$f(t) = 2\sin(18t) \cdot \sin t.$$

Compare with the figure at the end of Section 4.10.

87 LCR Circuit with Input Voltage.

(a) $\ddot{q} + 8\dot{q} + 25q = 150$.

(b) $i(t) = 50e^{-4t} \sin(3t)$,

(c) $\frac{d^2 i}{dt^2} + 8\frac{di}{dt} + 25i = 150\cos 3t$.

$$i_{st}(t) = \frac{75\sqrt{13}}{52} \sin(3t + \varphi) \quad \text{where } \varphi = \arctan\frac{2}{3}.$$

88 Planetary Orbits.

The two forces are equal: $GM = rv^2$.

Express v in terms of r and T.

7.5 Solutions Chapter 5

89 Euler's and Heun's Methods.
Euler: 1.10,   Heun: 1.10244,   exact: 1.1025

90 Model Problem.
(a) $1+h$,   (b) $1+h+\frac{h^2}{2}$,   (c) $1+h+\frac{h^2}{2}+\frac{h^3}{3!}+\frac{h^4}{4!}$.

92 Model for Rocket Launch with Air Drag.
Refinement of the problem solved in Section 3.8 by adding the air drag.

93 Model for Meteorite Orbits.
Data:
- Entry height $h_0 = 80$ km at time $t = 0$.
- Entry velocity v_0 relative to the atmosphere:
  12 km/s $\leq v_0 \leq$ 72 km/s.
- The weight G encompasses various orders of magnitudes:
  from 1 N, 10 N, 100 N to 500 kN (Hoba meteorite).

Cartesian coordinate system (x,h) where h = height, x = horizontal range.
Aim: To compute and represent the path of the meteorite and the velocity curve
as functions of time t.
Use the reasonings of Section 4.4:
The equation of motion with weight $\vec{G}$, drag force $\vec{W}$, and acceleration $\vec{a}$ is
$m \cdot \vec{a} = \vec{W} + \vec{G}$. The magnitude of the drag is given by

$$W(h) = \frac{1}{2}C_W \cdot A \cdot \rho(h) \cdot v(h)^2,$$

where A is the silhouette area and $v(h)$ is the speed.

As in (4.1), dividing the equation of motion by m and using $c(h) = \dfrac{C_W \cdot A \cdot \rho(h)}{2m}$

yields
$$\left. \begin{array}{rl} \ddot{x} = & -c(h) \cdot \sqrt{\dot{x}^2 + \dot{h}^2} \cdot \dot{x} \\ \ddot{h} = & -g - c(h) \cdot \sqrt{\dot{x}^2 + \dot{h}^2} \cdot \dot{h} \end{array} \right\}$$

Introducing the velocity components $u = \dot{x}, w = \dot{h}$ leads to a system of first order
differential equations in the four functions x, h, u, w:

$$\left. \begin{array}{rl} \dot{u} = & -c(h) \cdot \sqrt{u^2 + w^2} \cdot u \\ \dot{w} = & -g - c(h) \cdot \sqrt{u^2 + w^2} \cdot w \\ \dot{x} = & u \\ \dot{h} = & w \end{array} \right\}$$

Initial conditions:
$x(0) = 0$, $h(0) = h_0$, $u(0) = \cos\alpha \cdot v_0$, $w(0) = -\sin\alpha \cdot w_0$.
A further improvement could be to model the constant mass m by way of a
variable mass $m(h)$.

94 Trajectory of a Tennis Ball.

(b), (c): The trajectories are paths in space, as is often the case for cut balls.

95 Refined Predator-Prey Model.

(b) The solutions are no longer periodic.

As $t \to \infty$, they approach the stable fixed point $\left(\dfrac{c}{d}, \dfrac{a-kc/d}{b}\right)$.

96 Riccati Differential Equation.

(a) With $y_1 = 0$ one obtains $y = \dfrac{1}{\frac{b}{a} + De^{-at}}$.

Remark: With $y_1 = \dfrac{a}{b}$ one obtains the same set of solutions.

(b) $y = \dfrac{2}{t} + \dfrac{1}{kx^4 - \frac{1}{3}x} = \dfrac{2Cx^3 + 1}{x(Cx^3 - 1)}$.

7.6 Solutions Chapter 6

97 Climate Data.

(b) The horizontal line with mean surface temperature $16.7\,°C$ intersects the temperature curve at Pliocene level.

   If the c_2 values in the future do not exceed 491 ppm, the scenario would be near SSP1-2.6.

(c) In 2017 the ratio was 85.7% of that of 1987.

98 Carbon Budget for the 1.5 Degree Target.

$$c_2 = c_0 \cdot \exp\left(\frac{2 \cdot 1.35}{5.35}\right) = 463.8, \quad M_{em} = 671.3 \text{ Gt}, \quad \text{time period } 13.1 \text{ a}.$$

The probability that the effective temperature rise will be less than $1.5\,°C$ is only 50 %.

Note: Increasing the probability to about 66 % would almost halve the time span!

102 SEIR Model.

(a) $R_\infty = N - S_\infty, \quad I_\infty = E_\infty = 0,$

(b) $S_\infty \approx 0.2\,N, \quad R_\infty \approx 0.8\,N, \quad J_{max} = 0.153\,N$

(c) $0.91^x = \frac{1}{2}, x = 7.35$, mean infection period $= 20.4$ d.

   $$\frac{56\,d}{20.4\,d} = 2.75 \text{ infection levels.}$$

   $$R^{2.75} = \tfrac{1}{2}, R \approx 0.78.$$

103 Second Wave.

(c) (i) Equally high risk of infection:

First Wave: $j_{max} = 2.15$ % and $\lambda = s_\infty = 0.630,$
Second Wave: $j_{max} = 0.630 \cdot 2.15\% = 1.36\%$ and $s_\infty = 0.390.$

(c) (ii) Greater risk of infection:

$$j_{max} = 0.63 - \frac{1}{2.5}(1 + \ln(2.5 \cdot 0.63)) = 4.82\%.$$

This second wave is $4.82/1.36 \approx 3.6$ times stronger (higher) than in Case (i)!
$s_\infty = 0.234$ as a solution to the equation $s_\infty = 0.63 \exp[2.5(s_\infty - 0.63)].$

(d) The graphs intersect at $x = \lambda$ and $x = 1$. Since the graph of the exponential function $f(x)$ is convex, $f'(\lambda) < 1$. Also show that $f'(\lambda) = \lambda \cdot R_0$.

References

1. an der Heiden M, Buchholz U: Modellierung von Beispielszenarien der SARS-CoV-2-Epidemie 2020 in Deutschland. — DOI 10.25646/6571.2, Robert Koch Institut.
2. Blanchard, P., Devaney, R.L., Hall, G.R.: Differential Equations. Brooks/Cole Publishing Comp., Pacific Grove, California (1998)
3. Bogolyubov, N.N., Mitropolsky, J.A.: Asymptotic Methods in the Theory of Non-linear Oscillations. Gordon and Breach, New York (1961)
4. Boyce W.E., DiPrima R.C., Meade D.B.: *Elementary Differential Equations and Boundary Value Problems*, 11th ed., Wiley, 2017
5. Braun, M.: Differential Equations and Their Applications. 4th ed., Springer (1993)
6. Climate Change 2013: The Physical Science Basis. Fifth Assessment Report of the Intergovernmental Panel on Climate Change (IPCC).
7. Damon, P.E. et al: Radiocarbon Dating of the Shroud of Turin. Nature **337**, Nr. 6208, p. 611–615 (February 1989)
8. Engel, A.: Wahrscheinlichkeitsrechnung und Statistik, Band 2, p. 201–203, Klett, Stuttgart (1992)
9. Fässler, A.: Variationsrechnung mit Einführung in die Methode der finiten Elemente. Publikation Nr. 4 der Ingenieurschule Biel, nun Berner Fachhochschule, Hochschule für Technik und Informatik (1995)
10. Fässler, A.: Mittelungsmethode nach Bogoljubov-Mitropolski. Diplomarbeit am Institut für Angewandte Mathematik ETH Zürich (1973)
11. Fässler, A., Stiefel, E.: Group Theoretical Methods and Their Applications. Birkhäuser, Boston (1992)
12. Fässler, O.: C14-Altersbestimmung. Maturarbeit. Advisers: Irka Hajdas, Institute for Particle Physics ETH and Martin Lehner, Gymnasium Biel–Seeland (2010)
13. Garrigos, R.: De la Physique avec Mathematica. De Boek, Bruxelles (2009)
14. Golubitsky, M., Stewart, I.: Fearful Symmetry: Is God a Geometer? Dover Publications, New York (2011)
15. Goosse, H., Barriat, P.Y., Lefebvre, W., Loutre, M.F., Zunz, V.: Introduction to Climate Dynamics and Climate Modelling (2008–2010). Online textbook available at http://www.climate.be/textbook.
16. Graf, C., Ausbreitung von Epidemien, Maturarbeit an der Kantonsschule Büelrain, Winterthur (2017).
17. Hajdas, I.: Application of Radiocarbon Dating Method. Radiocarbon **51**, p. 79–90 (2009)
18. Heuser, H.: Gewöhnliche Differentialgleichungen, Einführung in Lehre und Gebrauch. Teubner, Stuttgart (1989)
19. Jazwinski, A.,H.: Stochastic Processes and Filtering Theory. Academic Press (1970)
20. Klvaňa, F.: Trajectory of a Spinning Tennis Ball. In: Gander, W. and Hřebiček, J, ed., Solving Problems in Scientific Computing using Maple and Matlab, 4th edition, p. 27–35, Springer (2004)
21. Leppäranta, M.: A Review of Analytical Models of Sea-Ice Growth. Online journal: Atmosphere-Ocean. Taylor & Francis, London (1993), Link to the article: http://dx.doi.org/10.1080/07055900.1993.9649465
22. Levin, I., T. Naegler, B. Kromer, M. Diehl, R. J. Francey, A. J. Gomez-Pelaez, L. P. Steele, D. Wagenbach, R. Weller, and D. E. Worthy (2010), Observations and modelling of the global distribution and long-term trend of atmospheric 14CO2,Tellus, Ser. B Chem. Phys. Meteorol., 62(1), 26–46, doi:10.1111/j.1600-0889.2009.00446.x.
23. Maeder, R.: Programming in Mathematica. Addison-Wesley, Publishing Company (1990)
24. Manabe, Syukuro, Broccoli, A.: Beyond Global Warming, Princeton University Press (2019)
25. Martin, J.E.: Mid-Latitude Atmospheric Dynamics. In: Subsection 2.2.2. Wiley, New Jersey (2009)

A. Fässler, *Fast Track to Differential Equations*,
https://doi.org/10.1007/978-3-030-83450-0

26. Maurer, P.: A Rose is a Rose. American Mathematical Monthly, August/September (1987)
27. Meinshausen, M.et al.: Geoscientific Model Development (European Geoscience Union EGU), The Shared Socio-economic Pathway (SSP) Greenhouse gas Concentrations and their Extensions to 2500, Geosci. Model Dev., 13, 3571?3605, 2020.
    https://gmd.copernicus.org/articles/13/3571/2020/
28. Metzler, W.: Dynamische Systeme in der Ökologie. Teubner, Stuttgart (1987)
29. Millington, J.: Curve Stitching. Tarquin Publications, Norfolk England (2001)
30. Murray, J. D.: Mathematical Biology: I. An Introduction, Third Edition, Springer Berlin, Heidelberg, New York (2002)
31. Paetsch, M.: Meteorit Neuschwanstein gibt Rätsel auf. Spiegel Online (8. Mai 2003)
32. Pólya, G.: How To Solve It, Princeton Science Library (2014)
33. Schneider, P.: Extragalactic Astronomy and Cosmology: An Introduction. 2nd ed. Springer, Heidelberg, New York, Dordrecht, London (2015)
34. Singh, S.: Fermat's Last Theorem. Harper Collins, UK (2012)
35. Tietze, J.: Einführung in die angewandte Wirtschaftsmathematik. 11. Auflage, Vieweg, Wiesbaden (2003)
36. Tierney, Jessica E. et al.: Past Climate Change inform our Future, Science **370**, eaay3701, 2020.
    Read full articel at: https://doi.org/10.1126/science.aay370.
37. Wagon, S.: Mathematica in Action: Problem-Solving Through Visualization and Computation. 3rd ed., pp. 578 +xi, Springer, New York (2010)
38. Wussing, H.: 6000 Jahre Mathematik. Springer (2008)

Internet:

39. Belk, J.: Illustration of the integral test in calculus.
    https://commons.wikimedia.org/w/index.php?curid=9786989. Public domain.
    Accessed in August 2017.
40. Stellungsnahme der Deutschen Gesellschaft für Epidemiologie (DGepi) zur Verbreitung des neuen Coronavirus (SARS-CoV-2)
    https://www.dgepi.de/assets/Stellungnahmen/Stellungnahme2020CoronaDGEpi-21032020.pdf
41. Gasser, H.-H., FIS-Sprungschanzen Baunorm 2018.
    Accessed in February 2020.
42. European Environment Agency: Atmospheric Greenhouse Gas Concentrations, 2020.
    Accessed in May 2021.
    https://www.eea.europa.eu/data-and-maps/indicators/atmospheric-greenhouse-gas-concentrations-7/assessment
43. Special Report: Global Warming if 1.5°C, IPCC, 2018.
    https://www.ipcc.ch/sr15/chapter/spm/
    Accessed in May 2021.
44. Lindsey, Rebecca: Climate Change: Atmospheric Carbon Dioxide, NOAA Climate.gov, Science and Information for a climate-smart nation, 2020.
    https://www.climate.gov/news-features/understanding-climate/climate-change-atmospheric-carbon-dioxide
45. Paleoclimate Working Group of the National Center for Atmospheric Research NCAR in Boulder, Colorado, supported by the National Science Foundation NSF.
    https://www.cesm.ucar.edu/working groups/Paleo/
46. Weisstein, E.W., Maurer Rose. From MathWorld — A Wolfram Web Resource.
    http://mathworld.wolfram.com/MaurerRose.html. Accessed in May 2017.
47. WolframAlpha, Computational Knowledge Machine. Free access for entering Mathematica commands. Accessed in August 2017.
48. https://de.wikipedia.org/wiki/Brüsselator. Accessed in June 2017.
49. Levin, I.: http://www.iup.uni-heidelberg.de/institut/forschung/groups/kk/en/14CO2_html
50. Graphs of Keeling curves in https://www.carbonbrief.org/analysis-what-impact-will-the-coronavirus-pandemic-have-on-atmospheric-co2

Index

Printed in the United States
by Baker & Taylor Publisher Services